A Beginner's Guide to Teaching Mathematics in the Undergraduate Classroom

This practical, engaging book explores the fundamentals of pedagogy and the unique challenges of teaching undergraduate mathematics not commonly addressed in most education literature.

Professor and mathematician, Suzanne Kelton offers a straightforward framework for new faculty and graduate students to establish their individual preferences for course policy and content exposition, while alerting them to potential pitfalls. The book discusses the running of day-to-day class meetings and offers specific strategies to improve learning and retention, as well as concrete examples and effective tools for class discussion that draw from a variety of commonly taught undergraduate mathematics courses. Kelton also offers readers a structured approach to evaluating and honing their own teaching skills, as well as utilizing peer and student evaluations.

Offering an engaging and clearly written approach designed specifically for mathematicians, *A Beginner's Guide to Teaching Mathematics in the Undergraduate Classroom* offers an artful introduction to teaching undergraduate mathematics in universities and community colleges. This text will be useful for new instructors, faculty, and graduate teaching assistants alike.

Suzanne Kelton is Associate Professor of Mathematics at Assumption University in Worcester, MA. She received her Ph.D. in Mathematics at the University of Virginia and has taught mathematics to learners across a range of universities and colleges.

A Beginner's Guide to Teaching Mathematics in the Undergraduate Classroom

SUZANNE KELTON

NEW YORK AND LONDON

First published 2021
by Routledge
52 Vanderbilt Avenue, New York, NY 10017

and by Routledge
2 Park Square, Milton Park, Abingdon, Oxon, OX14 4RN

Routledge is an imprint of the Taylor & Francis Group, an informa business

© 2021 Taylor & Francis

The right of Suzanne Kelton to be identified as author of this work has been asserted by her in accordance with sections 77 and 78 of the Copyright, Designs and Patents Act 1988.

All rights reserved. No part of this book may be reprinted or reproduced or utilised in any form or by any electronic, mechanical, or other means, now known or hereafter invented, including photocopying and recording, or in any information storage or retrieval system, without permission in writing from the publishers.

Trademark notice: Product or corporate names may be trademarks or registered trademarks, and are used only for identification and explanation without intent to infringe.

Library of Congress Cataloging-in-Publication Data
Names: Kelton, Suzanne, author.
Title: A beginner's guide to teaching mathematics in the undergraduate classroom / Suzanne Kelton.
Description: New York : Routledge, 2021. | Includes bibliographical references and index.
Identifiers: LCCN 2020029011 | ISBN 9780367429010 (hardback) | ISBN 9780367429027 (paperback) | ISBN 9781003000044 (ebook)
Subjects: LCSH: Mathematics–Study and teaching (Higher) | Mathematics.
Classification: LCC QA11.2 .K46 2021 | DDC 510.71/1–dc23
LC record available at https://lccn.loc.gov/2020029011

ISBN: 978-0-367-42901-0 (hbk)
ISBN: 978-0-367-42902-7 (pbk)
ISBN: 978-1-003-00004-4 (ebk)

Typeset in Avenir and Dante
by KnowledgeWorks Global Ltd.

Access the Support Material: www.routledge.com/9780367429027

To my husband, for his steadfast love and support.

To my son, for his positive outlook and
unrelenting belief in me.

To my daughter, for never letting me
forget what is truly important.

To Timothy, Alexander, & Rosabel
~ my rock, my light, & my love ~

Contents

Acknowledgements	ix
Introduction	1
Course Overview One	**6**
A Note on Terminology	7
Determining Desired Learning Outcomes	7
Assessment	13
Determining a Beginning Classroom Strategy	15
Backward Design	15
Quick Glance: Terminology Overview	16
Course Policies, Philosophies, and Syllabi Two	**17**
Course Policies	18
Syllabi	36
The Basics of the Classroom Three	**47**
The First Day of Class	47
Preparing for Class	53
During Class and Office Hours	63
After Class	85
Assessment Four	**87**
Fostering Academic Honesty with Assessment	88
Homework	90

Quizzes	90
Preparing Students to Take an Exam	92
Writing an Exam	97
The Exam Was Too Long!	100
Grading an Exam	102
Addressing Instances of Academic Dishonesty	106
Course Grades	107

Challenges and Opportunities within Commonly Taught Courses **Five** 111

Discussion Sections	111
Algebra – Is It Too Late?!	113
Precalculus	115
Calculus I: Differential Calculus	126
Calculus II: Integral Calculus	134
Sophomore Calculus	150
Elementary Linear Algebra	152
Proof Courses	156
Upper-Level Courses	159

Growth through Evaluation and Education **Six** 161

Self-Evaluation	163
Peer Evaluations and Collaborations	170
Student Evaluations	173
Pedagogical Professional Development	178

Going beyond Traditional Lecture **Seven** 179

Preparing to Try Something New	179
Active Learning	181
Course Designs for Active Learning	188
Online Courses	194
A Final Note on Innovation	197

Conclusion	199
References	200
Index	205

Acknowledgements

This book would not exist if not for the support of my colleague, James M. Lang. He saw potential in a scrappy little manuscript and offered the encouragement, advice, and guidance necessary to create the book before you. Thank you, Jim, for helping me make this book a reality.

My colleague, Charles Brusard, was the one of first to read the early manuscript and he urged me to pursue its publication. His ardent support was the driving force to begin the long journey towards the current book.

I extend my gratitude to Edward G. Dunne, who was Senior Editor of the American Mathematical Society (AMS) Book Program when I approached him in 2010 regarding the original manuscript. He generously pursued its posting on the AMS website, which has been a great honor.

Assumption University (then Assumption College) provided the sabbatical required to significantly expand and update the manuscript. Thank you to the staff at the Assumption University library for their efforts in retrieving research articles on my behalf.

I offer my sincere appreciation to those who provided essential feedback and invaluable suggestions for improvement through formal or informal reviews, especially Rachel Arnold, Jessica A. de la Cruz, Eric M. Howe, Steven Klee, and Fatemeh Mohammadi.

Special thanks to Tim Woodcock for reading countless revisions and providing assurances throughout this lengthy project.

Introduction

This guide is intended to assist a mathematician who has little or no teaching experience at the undergraduate level, but who will be teaching courses as a graduate teaching assistant or as a newly hired instructor. I will not tell you to get a good night's sleep or use the restroom before your first class, which is the sort of advice I received as a graduate student. Instead, I will review the basics of the classroom and course policy that any instructor should consider before the first day of class, as well as the running of day-to-day operations. In particular, this book addresses the specifics of mathematics instruction and its unique challenges. My aim is to present practical tools for achieving the goals you may already have for your classroom and assist you in determining your objectives if you are not sure where to start.

I am currently an associate professor at Assumption University, located in Worcester, Massachusetts. This book grew from my reflections on my teaching experience as I began preparing my teaching statement for my tenure dossier. The manuscript which served as a basis for this book was posted on the *Advice for New Ph.D.'s* page of the American Mathematical Society website in 2010 under the title *An Introduction to Teaching Mathematics at the College Level* and remains there at the time of this printing. I have significantly updated and expanded the original manuscript to include peer-reviewed research and updated strategies, reflecting my continued growth in my post-tenure career.

This book is structured to be useful if obtained just shortly before your first class meeting and remain a resource for courses you teach a few years from now as you craft and solidify your own philosophy. It begins with the essentials that you should consider before you step into class and offers more in-depth discussion in subsequent chapters. While a new instructor would benefit from reading and reflecting on the entire book prior to the start of class, the first three chapters would be the most essential. The following three chapters would preferably be read no later than the end of the first or second week of class. The final chapter could be left for later reading but the sections *The Art of Telling* and *Additional Methods to Achieve Active Learning* may be of immediate interest.

Chapter 1 asks you to start the process of determining goals for your course and how you will assess and foster learning in your classroom, illustrated through the example of a differential calculus course. While it is presumed that the primary audience for this book will not be tasked with designing a course, the latitude given will vary and thus an overview of potential considerations is provided. The foundational ideas introduced here will be explored more fully in subsequent chapters.

In Chapter 2, I discuss common course policies that any instructor should consider prior to the start of class. The chapter reviews the thought process behind the construction of a syllabus, from the basic, required components to additions that may help eliminate misunderstandings between professor and student. The topics covered aim to stimulate thought on the desired structure of the upcoming course and assist in a smooth start to the term. Pro/con tables of common options for assessment and grading illustrate benefits and pitfalls a new instructor will want to consider. A sample syllabus is provided to demonstrate how the suggestions may be utilized.

Chapter 3 highlights the general teaching practices that may apply for any course. The focus of this chapter is to encourage strategy and behaviors that will lead to a positive and successful classroom environment. The busy first day of class is addressed in addition to preparing for and conducting a typical class lesson. I discuss a variety of informal techniques to engage students and improve their retention and comprehension of the material. Methods for providing support and motivation to students are also examined.

In Chapter 4, I explore aspects of assessment. Specifically, I offer suggestions on how to help students prepare for a test, through the design of preceding in-class exercises, quizzes and your review guides, and the pros and cons of common test preparations are weighed. I outline the process of writing and grading an exam and options to consider if something goes awry. Your role in fostering academic honesty and addressing academic dishonesty are addressed. The chapter concludes with the responsibilities of assigning final course grades and weighing the role of extra credit in an undergraduate classroom.

Chapter 5 illustrates how the general advice offered in Chapter 3 can be harnessed in a variety of mathematics courses. In the lower-level courses, I discuss the types of choices an instructor might make in exposition and provide specific examples. I have included sample handouts which demonstrate how you can engage students in reviewing material and prepare them for testing. The sections for higher-level courses presume that some teaching experience is likely, so the focus narrows to specific opportunities which demonstrate how suggestions from earlier in the book can be executed within the course content. The practices of motivating discussion and drawing parallels to previously learned material to pique interest and improve learning are emphasized.

Chapter 6 proposes methods by which new instructors can continually evaluate and improve their teaching while simultaneously preparing to advance in their careers. Avenues to receive and process feedback are discussed, as well as how to document challenges and growth along the way. Several worksheets are included which offer a manner to efficiently collect and analyze data on your teaching. This documentation serves to guide the pursuit of improvement, in addition to recording the journey. Even readers who do not care to prepare a teaching portfolio, such as graduate students not intending to enter academia, can use these methods to ease and improve their teaching experiences in graduate school.

Chapter 7 provides an overview of some alternatives to the standard lecture. While a variety of active learning strategies are discussed in Chapter 3, this section lists additional methods and offers resources for further investigation. A few active learning course designs are also discussed in this chapter, including inquiry-based learning, community service learning, and flipped classrooms. These are not recommended in their entirety for a beginning instructor outside of a mentored program. As such, each is described here simply for the reader's exposure and is not explored in detail. The methods recommended earlier in the book and at the start of the chapter will provide a first step towards adopting a more active course design. A brief discussion of online courses and the making of videos for online courses is included.

While ideally, this book would be read in its entirety prior to the first day of class, time constraints may not permit the reader to do so. The following chart depicts the latest one might read each chapter, along with the core topics covered.

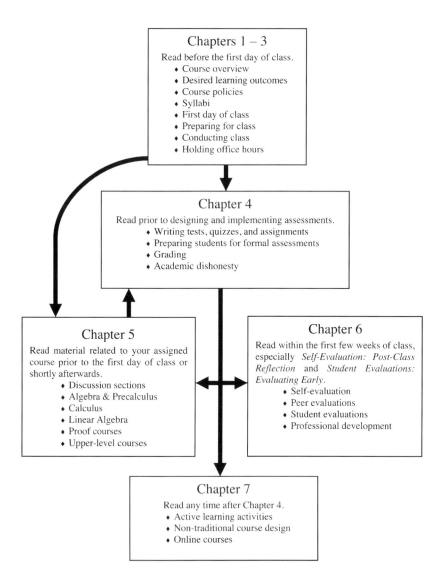

There are a number of companion materials available at www.routledge.com/9780367429027. In the book, the majority of the Chapter 5 course worksheets are shown complete with suggested responses. The worksheets, along with blank versions, and a few bonus handouts, are provided online. The evaluation and reflection worksheets from Chapter 6 are also posted there for your convenience.

Whether or not you are embarking on a career in academia, this book aims to make your immediate teaching experience enjoyable and productive. There are many interesting challenges ahead of you, but hopefully there are exciting discoveries and successes awaiting you as well. Remain open to new ideas and the suggestions of your colleagues throughout your journey. If you feel isolated and without a community at your current institution, be assured there is a global community available and additional resources will be suggested in this book.

Course Overview   1

Why are you about to teach a math class? You may be taking the first step towards a lifelong career in teaching or perhaps you are simply offsetting costs while in graduate school. Hopefully, you not only possess competency in the mathematics you are about to teach but also a love for math in general. Whether you will be teaching a lifetime or for just a few years, the classroom is an opportunity to share what you love about math, assist students' discoveries, and entice new enthusiasts to the field. As you approach the course ahead, consider what you want your students to learn beyond the obligatory skills. Where is the beauty or intrigue?

I have always been fascinated by the connectivity within mathematical material, especially when two concepts initially appear quite distinct. Perhaps it is no surprise that I was drawn to the field of topology, where two very different-looking surfaces can be revealed to be effectively the same and where we can learn about a perplexing space by turning to another we already understand well. As I will discuss in various points in the book, making connections between concepts by drawing parallels is an effective learning tool. Since I find it quite fun, this practice has been a valuable and enjoyable part of my classroom experience. Examples for specific courses are provided in Chapter 5.

In this chapter, I ask you to think about your overall learning goals for the course, as well as the particular skills your students should gain. Even if you have a very rigid course outline to follow, taking just a few minutes to look over the concepts and skills that will be covered and to reflect on how they intertwine is time well spent. While I will discuss how to articulate and assess the desired learning in your course, I have presumed that most readers of this book are not immediately tasked with fully designing a course.

Most new instructors will be given a list of the expected course content but the degree to which these are detailed will vary by institution. Instructors who have been given a wide latitude with their courses may not have much more than a catalog course description at their disposal. The material covered here is provided with these instructors in mind, but also so that all readers may understand the process behind specifying course goals and consider the nuances in those set by their department.

A Note on Terminology

There are several common terms for describing the specific and general learning aims you may have for your students. Unfortunately, terms such as *course goals*, *learning objectives*, and *learning outcomes* are often conflated and used interchangeably. When I discuss *course goals*, I am referring broadly to the general knowledge and skills students are intended to learn in the course. The set of *learning objectives* is a list of skills or abilities you want students to have at the end of the course. *Learning outcomes* are the specific ways that students will evidence how the learning objectives have been achieved. A learning objective might be demonstrated by a number of distinct learning outcomes, which provide insight as to the manner and depth to which an objective is realized. This will be illustrated in the next section through the context of a differential calculus course.

Determining Desired Learning Outcomes

Developing an accurate and realistic set of objectives is challenging the first time you teach a specific course. If you have never taught any undergraduate math before, you have most likely been given a sample (if not mandatory) syllabus for your course which depicts the material which you are expected to cover and a text should have been pre-selected or recommended. Seek input from the chair of the department and instructors who have recently taught this course, request prior syllabi if they have not been provided, and refer to your institution's course description. Previous instructors can provide valuable information about the essential skills needed in subsequent courses, as well as the typical student preparedness. Using the information you have obtained, list the objectives (i.e. desired skills or abilities) you have for your class.

Perhaps you have numerous skills you want students to master or maybe you have only been able to outline the general course goals. When you approach a new course, you may overestimate what every student *needs* to know in a few months' time. Any topic we might choose has many interesting avenues or applications that could be explored, more than any student could digest entirely in one semester. Conversely, there may be skills that you initially take for granted but later discover some of your students lack or find more challenging to learn than you expected. Perusing the text may help you gain a sense of the level at which the material will be discussed and the general approach to the course content. Again, reaching out to previous instructors for guidance can be quite helpful. Regardless of how complete you may feel your list may be, you will likely need to revise and refine as you move though the course the first time.

If you have only been able to list general course goals, consider what you want your students to learn from each of the topics included and more specifically, how will they demonstrate this? For instance, rather than saying you want students to *"understand* that the rate of change of a function is given by its derivative," what action students can perform that will illustrate comprehension of this concept? We might identify the learning objective as: "Students should be able to demonstrate knowledge of the relationship between a function and its derivative." One or more of the following learning outcomes could be used to measure the achievement of this objective, depending on the approach of the course:

Students should be able to:

- Find the rate of change of a function.
- State what the derivative of a function reveals about a function's behavior.
- Given the graph of a function, identify the graph of its derivative(s).
- Use the derivative to state the intervals where a function increases or decreases.
- Construct the graph of a function using its derivatives.

Actively framing your statements clarifies how you will assess the achievement of your objectives and may assist you in designing more enlightening examples and in-class activities.

Suppose you are tasked with a differential calculus course for non-majors and the course description states that topics include limits and continuity, derivatives of a variety of functions, and applications of differentiation. You might create the following list of learning objectives:

Students should be able to:

- Determine the best method for calculating the limit of a function and find the limit.
- Determine the continuity of a function at a point.
- Use the formal definition of the derivative and explain its relationship to rate of change.
- Describe the relationship between continuity and differentiability.
- Demonstrate an understanding of the relationship between a function and its derivatives in concrete and abstract settings.
- Compute the derivative(s) of a function, by identifying and executing the necessary differentiation rules.
- Identify and complete applications of the derivative.

Here is a sample list of learning outcomes, indicating how students would demonstrate how the above learning objectives were achieved:

Students should be able to:

- Find the limit of a function as the input approaches a finite value using a table of values, limit laws, the Replacement Theorem, and by inspection of its graph.
- Find limits at infinity using algebraic methods, as well as by inspection of a graph.
- Identify discontinuities of a function using its algebraic definition, as well as by inspection of its graph.
- State the definition of the derivative and use it to compute the derivative of polynomials, rational functions, and radical functions.
- State what the derivative of a function tells us about the function's behavior.
- Compute derivatives of polynomials and exponential, logarithmic, and trigonometric functions using differentiation rules.
- Compute the derivative of products, quotients, and compositions of polynomials, radicals, and exponential, logarithmic, and trigonometric functions.
- Use derivatives to find critical numbers and points of inflection for the graph of a function.

> - Use derivatives to determine extrema and concavity for the graph of a function.
> - Use derivatives in applications, such as distance/velocity and the construction of graphs.

Suppose instead that you will be conducting a more advanced differential calculus course for math majors. Your learning outcomes would naturally look different despite the same general subject matter of the course. Your list might grow to include the following tasks:

> - Find limits at infinity using L'Hôpital's Rule.
> - Use limits to determine the continuity of a function at a given input value.
> - Use limits to compute derivatives of select trigonometric functions.
> - Use the derivative to compute the rate of change of a function.
> - State and prove the Mean Value Theorem and be able to use it in the context of average rate of change.
> - State and prove Rolle's Theorem.
> - Compute derivatives of implicitly defined functions.
> - Use derivatives to solve optimization application problems.
> - Describe and construct graphs using derivatives.

Courses intended for science or engineering majors would most likely present a shift towards applications in those fields. If you are teaching a highly coordinated course, then you may have received a syllabus with a detailed list of desired learning outcomes or objectives complete with a daily schedule of the material to be covered and homework assignments. Even though you might have little freedom in a course such as this, reviewing the details provided allows you to understand the depth and scope of the course.

Initially, you might list mainly computational skills for the desired learning outcomes in lower-level courses while reserving more conceptual material for more advanced courses, but you will want to strive for a blend of both whenever possible. One way to keep a check on the balance within your course, and to potentially expand your initial goals, is to consider L. Dee Fink's Taxonomy of Significant Learning. The idea behind this taxonomy is that the most significant learning occurs when you can achieve all these various aspects in your presentations, discussions, and overall approach to the material. Fink describes the following six categories (Fink, 2003: 30–2):

L. Dee Fink's Taxonomy of Significant Learning

Category	Description	Examples or Opportunities in Calculus
Foundational Knowledge	A basic understanding and recollection of primary facts or skills	Limit laws, a graphical understanding of continuity, the definition of a function, and derivative rules
Application	The ability to use the foundational knowledge and apply critical thinking	Using limit techniques, computing derivatives of complex functions, and using derivatives to determine extrema and concavity
Integration	Connecting ideas or topics	The connection between a function's graphical representation and that of its derivatives
		Connections with the skills to those used in another course, such as chemistry or physics
		The use of Riemann sums to illustrate the discrete setting, such as in finding area and volume, and the connection to the continuous case
		[Some courses offer ample opportunities to draw connections to prerequisite material. For instance, you might draw a parallel to connecting polynomial long division to numerical long division.]
Human Dimension	Personal and social implications of the material	Applications for health sciences, such as rate of disease spread or concentration of medication in the bloodstream
Caring	An increase or change in caring about a topic	Applications targeted at majors in your course, a community service project
		Having students discover the need for a new technique or skill, such as dy integration, because the prior one is ineffective or tedious
		Demonstrating your own enthusiasm for the material and why you find it interesting

(continued)

(continued)

Category	Description	Examples or Opportunities in Calculus
Learning How to Learn	Learning about the process of learning	Asking students to make predictions about the correct strategy for a problem and subsequently analyzing why they made their predictions and why those predictions were correct or incorrect Asking students to create examples or counterexamples

True comprehension of foundational knowledge is achieved when students gain a command of the conceptual structure of the material (Fink, 2003: 36), which we can achieve, in part, by applying the foundational knowledge and connecting new concepts to what we have learned previously. In Chapter 5, I discuss a variety of opportunities to draw parallels and connect material within the specific courses addressed. Students are often quite pleased by this interconnectivity of the material and form a new appreciation for the mathematics they are learning, thereby providing one avenue to caring.

We can use the taxonomy to more fully develop to the first set of learning outcomes created. *Foundational Knowledge* and *Application* are well represented in the original list but we could add the following to address *Integration*:

- Discuss the connection between continuity and differentiability.
- Discuss the connection between the graph of a function and the graph of its derivatives.

To address *Human Dimension* and *Caring*, we could include:

- State at least two uses of derivatives in the real world.

While you may ultimately infuse *Learning How to Learn* mostly in your classroom strategy, you could include:

- Describe the process by which a function is analyzed to determine which derivative rules are necessary in computing its derivative.
- Construct examples, such as that of a function which will satisfy a set of conditions on its derivatives or its graph (asymptotes, discontinuities, etc.).
- Construct counterexamples to prove a statement is false, such as one involving the relationship between continuity and differentiability.
- Correct false statements.

Some courses will lend themselves better to drawing connections or demonstrating uses in the community, but keeping the full taxonomy in mind can help you bring the most to your classes.

Quick Glance: Tips for Determining Learning Outcomes

- Obtain prior syllabi for your course.
- Examine the textbook to assess the level, if one was preassigned or recommended.
- Confer with colleagues who have taught the course, or your department chair, to develop and review your list.
- Brainstorm specific demonstrative skills or concepts that you want your students to possess at the end of your course.
- Review the taxonomy and evaluate whether you have addressed each category.

Assessment

Once you have thought about the knowledge and skills your students should gain from your course, consider how these will be assessed. In Chapter 2, I will delve into the pros and cons of different formal assessments, such as tests, quizzes, and assignments. Here, I am asking you to consider the manner in which you want students to demonstrate knowledge and proficiency in your course. For instance, *how* will you ask students to discuss the derivative of a function? Will you expect students to analyze graphical representations, algebraic representations, or both? Will your assessments be highly computational or involve theoretical analysis and discussion?

The way you test your students should relate to how the material is discussed in class and in homework. In *What the Best College Teachers Do*, Ken Bain observes that "the best teachers see examinations as extensions of the kind of work that is already taking place in the course" and "[t]he goal is to establish congruity between the intellectual objectives of the course and those that the examination assesses" (2004: 162). If the test format that comes to mind is not in line with how you have thought about discussing content in class, consider how you might blend the two.

If your course is highly coordinated and you will not be writing the assignments, quizzes, or exams, seek guidance from the coordinator as to how students will be tested on material. The coordinator may be able to provide you with prior exams to illustrate the assessment strategy. Students can possess a solid understanding that the derivative gives the slope of a function and still struggle to translate that knowledge into the matching of the graph of a function to that of its derivative. Understanding the ways your department will evaluate your students' comprehension should contribute to the way you intend to discuss the material.

Varying the way you assess content provides increased opportunity for students to demonstrate knowledge and to do so in more depth. For instance, short computational exercises demonstrate basic competency while those which ask students to discuss a relationship allow for evaluation of conceptual understanding. Offering these different forms of evaluation serves to measure the different strengths your students possess and may even impact the occurrences of academic dishonesty (see *Fostering Academic Honesty with Assessment* in Chapter 4). Determining how you want to assess diverse aspects of a topic informs the desired learning outcomes for the course, so you may find you need to revise your original list after these considerations.

Informative, frequent assessment is important in any subject, but especially so in mathematics. Providing precise and prompt feedback which addresses the strengths and weaknesses of students' work is vital to their understanding how to improve (Mathematical Association of America (MAA), 2018: 56). Students must be alerted to misconceptions and errors early to avoid practicing incorrect methods. In *How Learning Works: Seven Research-based Principles for Smart Teaching*, Susan Ambrose and her co-authors note "more frequent feedback leads to more efficient learning because it helps students stay on track and address their errors before they become entrenched" (2010: 142). They assert that feedback should state where students stand in terms of the course goals and should be given when students can best harness it, in light of the goals and planned activities (2010: 138). If you have control over the assessment schedule, consider how to place assessment activities with the desired learning outcomes in mind. For instance, prior to embarking on complex derivatives or applications of derivatives, it is important to first assess students' abilities in the differentiation skills covered thus far and provide feedback on their errors. Contemplating the way skills in your course build may suggest additional desired learning outcomes.

Determining a Beginning Classroom Strategy

Your plans for classroom strategy in a first course are likely to be fairly broad. First, consider a general approach to the material. From the preceding discussion for a differential calculus course, a new instructor might strive to motivate discussion of the course topics graphically or with applications whenever possible, introduce basic skills in relation to the opening examples, then proceed to more complicated and abstract content. With this in mind, consider ideas for how you will conduct class meetings. You might envision students working together often and offering in-class quizzes or worksheets frequently to provide feedback and gauge students' comprehension. I will explore various components of strategy in the coming chapters to assist in creating a more concrete plan. Keep an open mind as you progress through new courses; understanding what works best for an individual course and its typical audience takes time and experience.

Backward Design

The approach I have outlined in this chapter follows the "backward design" in which the instructor first determines the desired skills, then how to assess these skills, and finally the ways in which the desired learning can be achieved through the design of the activities and instruction. This strives to avoid what the authors of *Understanding By Design* call the "'twin sins' of traditional design", using activities purely for engagement, rather than learning, and covering content with no indication of the educational purpose (Wiggins & McTighe, 2005: 16). In *Preparing for Class* (Chapter 3), I will revisit this approach when constructing the lesson(s) for a chosen topic. In that example, I include specific notes on the approach to the material and the interaction in the classroom.

While you may have ideas regarding strategies as you formulate your course learning objectives, it will be easier to design and articulate these after teaching the course at least once and gaining a more defined sense of the appropriate level of the content and the background of students. Instructors should remain flexible as they face a new course and learn its audience. It is necessary to formulate what you aim to achieve at the start but be open to tweaking the intended list of outcomes and objectives, while remaining faithful to any declarations of coverage laid out in the syllabus. Depending on the detail of your syllabus, you may have flexibility mid-course to adjust your desired learning outcomes. Take note of any goals, objectives, or outcomes which were unrealistic or insufficiently challenging and consider how you could revise them in the future.

Quick Glance: Terminology Overview

Topic	Definition	Purpose
Course goals	General knowledge and skills to be obtained in the course	Determines the content for the course, with an overarching approach in mind
Learning objectives	Specific, demonstrable knowledge and skills to be obtained in the course	Formulates the course goals actively
Learning outcomes	Observable actions which demonstrate the manner in which the learning objectives have been realized	Articulates the aspects of the material which are most valued and guides the methods of assessment
Assessment	A measurement of the degree to which a student's attempt at a learning outcome is successful	Provides feedback for students and instructors; determines the degree to which the learning objectives have been achieved
Backward design	An approach to planning which starts by identifying desired abilities, followed by determining how these will be evidenced in assessments, then how to design instruction and activities	Provides focus and direction for instruction and activities

2

Course Policies, Philosophies, and Syllabi

The first class I taught in graduate school was a coordinated differential calculus course for which all sections used the same departmental syllabus and homework assignments. While I was thrilled to learn, upon obtaining my first tenure-track position, that I would have full autonomy in my courses, the setting of coordinated classes is excellent for beginning instructors. If you are teaching the same course alongside others, who are all following the same course schedule and assignment list, you have the opportunity to observe, reflect, and discuss issues in real time. If, however, you find yourself at a small institution with little oversight or mentoring immediately available, you may have to put in quite a bit of effort to educate yourself on your new institution's practices and policies. If you have not been given a fully developed syllabus for your course, this chapter will guide you through the various components you will want to consider.

The syllabus is your contract with your class. It lays out the course content, grading structure, and the overall goals you have for your students. Your syllabus should not only discuss all aspects which will affect the course grade but also the course design and desired atmosphere in the classroom. It is important to check in with your institution to verify which material is required on their syllabi and learn the accepted practices for course policies. The course policies should be clear and you should abide by any declarative statements you make. If a component of the course is flexible (such as the total number of assignments to be given), then the syllabus should reflect that. My syllabi have evolved greatly over the years, starting with just covering the basics of the course such as topic coverage, text, and grade breakdown, to now discussing many of the policy issues that arise each semester.

Coordinated classes can vary in the degree to which the course components are aligned. Some will only take a common final, while highly coordinated classes will have a predetermined schedule and common syllabi, assignments, quizzes, and examinations, leaving virtually no administrative or structural decisions to make. The opportunity to see the department's course design and ultimate required skill set in the form of a common final is an excellent learning opportunity. Even when you disagree with how content is handled, you develop a sense of what you value most in a course which will be vital when the time comes to design your own. Similarly, as you read through this chapter, you can weigh the options even if you are not currently required (or allowed) to make any decisions presently.

Course Policies

Whether or not you will be tasked with creating a syllabus, you should consider the policies you want to implement in class and those expected by your institution. If you were given a syllabus to use, are any of the following topics not covered? You should check with a course coordinator or department chair to determine if there are official policies or standard practices at your institution. If there are issues such as how to handle missed quizzes that are left to your discretion, you may want to create a supplement to the department-provided syllabus that addresses those points. At the very least, you will want to consider how you will handle each, and possibly reach out to colleagues or superiors for ideas or feedback, before you step into class that first day.

Grades

If the decision about how the grades will be calculated is wholly in your control, there are many aspects to consider. You may still need to consult your institution's policies, such as any stipulations regarding the final exam or project. You may be required to have a cumulative final and there may be requirements about the weight of the final in the course grade. After satisfying any required elements, consider the remaining components. What types of formal assessment do you want to include – tests, quizzes, homework assignments, projects, presentations? How many exams do you want to give and how often do you want to administer quizzes and homework? How much should each be worth? What about participation? Extra credit? How would an honor violation affect a student's grade? How long might you allow a student to contest the grading of a paper or make up a missed assessment?

What to Grade?

You can start by considering different ways students may demonstrate the knowledge you seek them to possess. Try to provide avenues for a variety of levels of learning and abilities to be represented in your grading scheme (Fink, 2003: 142). Leaning too heavily on one assessment tool will provide a distinct advantage to students who excel with that format and potentially under assess the abilities of those who do not. For instance, if you give an in-class midterm and final exam as your only graded work, you will be assessing students' ability to handle roughly seven to fourteen weeks of material at once and to perform it in a limited amount of time. Allotting some portion of your grade for quizzes allows demonstration of short-term mastery of select skills. Including a homework or project grade provides the opportunity to show work on un-timed exercises.

Weekly quizzes are one way to achieve regular feedback and assessment and they allow students an opportunity to see how they will be tested prior to an exam. Frequent quizzes can serve to improve learning and retention of material (Brown, Roediger, & McDaniel, 2014: 226). If you cannot offer frequent graded quizzes due to time constraints or course coordination, you can utilize ungraded quizzes and other in-class exercises discussed in Chapter 3 to provide this essential feedback, as well as using methods discussed in *Additional Methods to Achieve Active Learning* in Chapter 7. Providing opportunities for students on quizzes (or in-class exercises) to work in a format similar to those used on examinations allows them to gain a better understanding of the depth and breadth to which their competency will be assessed. This will be addressed in more detail in Chapter 4.

Projects provide one path to assessing long-term skills, without the test-taking environment, but may be less advisable in your first run with a course. Selecting or designing projects that will be both an efficient use of your students' time and an effective learning tool requires a solid grasp of your learning objectives and a knowledge of the abilities and prerequisite skills of students typically enrolled in your course. This is not to say you should not attempt a project that seems like a good fit, but you might consider keeping it small for a trial run. Perhaps, you could break it down into smaller assignments or in-class activities. This allows you multiple opportunities to adjust and modify, if needed.

Keep in mind that with each additional piece of formal assessment comes more grading. Larger institutions may provide graders to ease this burden but smaller ones often offer little or no support in this way. Knowing that collecting frequent homework assignments and holding weekly quizzes provides frequent and more balanced feedback has to be weighed with the practical reality that these papers must be returned in a timely fashion – as soon as possible – if that feedback is to be of use. If you assign projects, can you grade these with the attention they deserve in the time you have? Additionally, if you have a number of students with learning disabilities, who need additional time on quizzes, will you be able to accommodate this every week? Possibly, your institution has a testing center or academic support center which can assist. Having more than a few students who require additional testing time outside of class is a concern you must address realistically.

One answer to quickly grading numerous homework assignments is the use of an online homework system. These bring their own issues of accepting answers in a different form than is sufficient for your quizzes and tests or not accepting answers due to minor deficiencies. The lack of the computer grading system to give detailed feedback, in noting where the error occurred in a student's work, or to give partial credit, limit the effectiveness of this format to acknowledge what the student has done well and inform on what needs work.

Any time you create an assessment, give frank consideration to what you will be measuring. Using homework as an assessment is challenging and not simply from the grading aspect highlighted above. It is *imperative* that students complete basic exercises to practice and learn techniques, but a graded assignment consisting of such problems may assess very little. Many problems can often be completed in seconds with an internet search, online calculator, or app. This holds true for an extensive array of math courses, even those involving proofs. There are also subscription services available which provide solutions to both odd and even problems in thousands of texts. Such resources provide detailed steps, so requiring students to show their work does not necessarily demonstrate their effort or *any* understanding of the process. If you want a homework grade to assess comprehension and effort, it will require work on your part to design unique assignments containing questions such as true/false with a justification, short response, and applications. This is not to suggest that basic exercises should be eliminated! These are an absolute must for students to learn and practice skills. My comments here are intended to highlight the issues that arise in *grading* such problems and considering them an assessment of skill and effort.

While effort can be difficult to assess formally, students appreciate when some portion of the course grade attempts to do so. Keeping in mind the above concerns for the design of graded assignments, a portion allotted to homework for which students are allowed to collaborate and/or seek help from you or a tutor is perhaps the simplest way to try to reward hard work since each student has the opportunity to produce an accurate and complete paper. Another avenue is a participation grade but this can be challenging for mid-level students who do not know an answer immediately in class, but are solid enough to complete most of their work without assistance. If you have students work in pairs or groups during class, this may allow you to better discern the engagement of those students and provide them a better opportunity to contribute.

Pros and Cons of a Selection of Formal Assessment Tools

Form of Assessment	Pros	Cons	Things to Consider
Exams (in-class)	Assess the level of mastery on an array of skills and difficulty levels Assess the ability of students to complete work on their own	Only assess students' ability to perform skills in a timed environment Students may not be receiving sufficient and timely feedback if this is the only method of assessment utilized.	While less frequent than quizzes, exams will be longer and more complex, requiring significant time to grade. Students receiving accommodations for disabilities may require additional time outside of class.
Exams (take-home)	Permits testing with in-depth, complex problems Provides extended time for testing	Security is a concern, as it is easy for students to collaborate or seek answers from unauthorized sources. Students may not be receiving sufficient and timely feedback if this is the only method of assessment utilized.	While less frequent than quizzes, exams will be longer and more complex, requiring significant time to grade.

(continued)

(continued)

Form of Assessment	Pros	Cons	Things to Consider
Quizzes (in-class)	Provide timely and focused feedback Only test a small amount of information at a time	Only assesses students' ability to perform skills in a timed environment Creates a volume of papers to be graded routinely – and quickly	Prompt grading is necessary and could pose a challenge. Students receiving accommodations for disabilities may require additional time outside of class.
Homework	Provides timely and focused feedback – if you are doing more than a check for completeness Only tests a limited amount of information at a time Allows students to demonstrate abilities in un-timed setting	May be difficult to discern whether or not students have completed the work themselves and/or understand the work they have submitted	May be burdensome to complete grading in a timely fashion, which is necessary for proper feedback A check for completeness does not provide feedback and may not differentiate student effort.
Projects	Allows for un-timed work Provides an opportunity for in-depth engagement with the material which may connect various course topics Provides an opportunity for developing problem-solving skills Group projects may teach teamwork and valuable strategies for the workplace.	May not provide sufficient timely feedback on essential skills May not provide the opportunity to demonstrate abilities in basic skills if students struggle with the complexities of the problem Grading may require significant time. Group projects may lead to issues with teams which are unbalanced, do not work well together, or have members not contributing to the effort.	Effective project design requires considerable time and a deep understanding of the course objectives. May be burdensome to complete grading in a timely fashion If assigning group work, effort must be made to create balanced groups and difficulties within groups may arise.

(continued)

Form of Assessment	Pros	Cons	Things to Consider
In-class group work (as a measure of participation)	Supports peer interaction. Allows students to engage during class. May allow you to observe students who do not otherwise speak up in class	Students may need frequent redirection to stay on task. Task may take more time than originally intended, depending on the level of student engagement. (This is only a con if there is low engagement!)	May be difficult to create a meaningful rubric to assign individual grades in a clear, consistent, and transparent manner

How to Calculate Grades?

After you have determined the forms of assessment you wish to utilize, consider the weight of each. Suppose you have decided to use quizzes, homework, two midterms, and a final exam. Most simply, you could weight these each 20%. Perhaps you want examinations to count for more than the other components. You could also opt to have each of your exams weighted differently.

A weighting method to *avoid* is one based on performance, as this can fail to preserve class rank. For instance, suppose a course has three mid-semester exams and a student's best exam counts as 30%, the next highest as 20%, and the lowest exam as 10%, with the remainder of the grade to be determined by other assessments, including a comprehensive final. This approach might appeal to an instructor who wants to lessen the impact of a single exam but it can fail to accurately represent the overall class rank of students. For example, suppose student A earns an 80% on each test, while student B earns 100% on the first, 80% on the second, and 50% on the third. Student A has performed consistently throughout the course, while student B's work has steadily deteriorated. If the grades were weighted evenly, student A would have an 80% test average, while student B's average would be a 76.7%. Using the 30-20-10 weighting plan described above, A's test average still be 80%, whereas B's would be 85%. The good intentions of the professor to give student B a break has unfairly boosted this student above student A. This is especially unjust if the material on the third test was more complex and sophisticated than that which was on the first test.

A *fair* alternative to evenly weighted exams is to have weights increase throughout the semester as additional or more complex material is added. In Chapter 4, I will discuss the practice of including small cumulative sections to all examinations. In this setting, it might be appropriate to have the first exam count as 15%, the second as 20%, and the third as 25%. The earlier exam covers less material, so it is reasonable that it does not count as much towards the course grade. This weighting structure preserves class rank but lessens the impact of a poor performance on the first exam. This can work for or against students, depending on why they struggled and whether they take the feedback seriously. Students who recognize deficiencies in study habits and adjust accordingly can recover nicely, but some might not have prepared as seriously for a low-weighted exam. This decreases your ability to accurately assess their skills and lessens the effective feedback they receive. Those who are ill-prepared for the course may not take feedback from the first test as seriously as they should since the overall grade has not yet been severely impacted. While most math courses do tend to build on earlier material, some courses are more disjointed. This weighting format is difficult to justify in a course that does not have a cumulative component in the testing, unless certain material has obvious significance. There may be other justifications for giving greater weight to the material on certain exams but it should not be done arbitrarily to lessen the impact of low grades on assessments.

Pros and Cons of Different Weighting Schemes

Weights	Example	Pros	Cons
All instances of a single assessment type weighted evenly	Each exam is weighted 20% in final course grade.	Assesses the level of mastery evenly throughout the term Preserves class rank Gives equal importance to each exam Students may place significance on early feedback.	Students who do not fare well on an early exam may withdraw, even if better study habits and test preparation may be all that is necessary to improve performance. A single poor performance may have a significant impact.

(continued)

Weights	Example	Pros	Cons
Instances of a single assessment type weighted by performance	Highest exam grade is weighted 30%, second highest weighted 20%, and lowest weighted 10%.	A single poor performance may not significantly impact the course grade.	Does not assesses the level of mastery evenly throughout the term
			May fail to preserve class rank
			Does not give importance to any particular exam, potentially impacting student preparation
			Students may fail to place necessary significance on early feedback.
			Difficult to assess students' standing until all tests have been administered
Instances of a single assessment type weighted by significance or amount of content covered	Exams containing at least some cumulative material increasing in weight, such as first 15%, second 20%, and third 25%	A poor performance on the first test has a lowered, but fixed, impact on course grade.	Decreased weight to some assessments may impact student preparation.
		If the tests are cumulative, students may better attend to early feedback, despite the lowered effect on course grade.	Students may fail to place necessary significance on early feedback.

Flexibility

Consider giving yourself a little latitude in the precise number of the less impactful graded components of your course, such as quizzes and homework assignments. Suppose you plan to give ten quizzes and want quizzes to comprise 20% of the course grade. You may find it easier to make the quiz *average* worth 20% of the grade rather than planning ten quizzes, each 2% of the final grade. This way, you can avoid headaches if cancelled classes or other delays interfere with your plan to have exactly ten quizzes. Alternatively, you could state a schedule which gives a sense of the number of expected quizzes. For instance, you could state that there will *typically* be a quiz each week, except for test weeks. Whatever your approach, you should give students a good idea of the approximate number of graded papers, so they understand roughly how much weight they carry individually.

Your first time with a course or at a new institution, you may be unsure whether you will be able to include all of the components you have considered. Suppose you have a small project in mind but are concerned there may not be sufficient time or that it may not be a good match for the audience. You can address the possibility by indicating that assignments may include a small project(s). In this case, you can clarify the weight of any assigned project, such as two or three homework sets. This alerts the class to the possibility of a project and how it would factor into the grade but does not commit you to including one.

While the formal assessment tools and the impact each will have on the final course grade must be determined prior to your first day of class, Chapter 3 will discuss informal options. These impactful activities are aimed at improving your students' comprehension and retention, and thus their performance. Since these are informal, they can be added to increase your students' learning and potentially improve upon your initial plan of assessment without violating your syllabus.

Participation and Attendance

In many cases, a student's participation and attendance affect their final course grade whether or not it is a formal part of the grade. If you want either of these to formally factor into your grade calculation, consider how you would assess each. This should be clearly indicated in the syllabus to avoid any misunderstandings.

In the case of participation, you may want to respect the shyness a student may feel towards speaking out in class. A student who does not know the answer to a question on the spot, nor is able to immediately formulate a question, is not necessarily removed from the discussion. This can be especially true for students with learning disabilities, who may require recording a lecture or having another student take notes for review at a later time. One way to address these issues is to allow asking questions in private settings, such as during office hours or through e-mail, to count towards participation in the course. I try to make the point on day one and throughout the semester that *routine* attendance in my office hours contributes to a student's participation grade as one way to encourage students to stop by.

A student's attendance is much easier to quantify but it may not be necessary to include it in your grade calculation directly. Students who frequently skip classes usually do not do particularly well, especially if the lectures, class discussions, and in-class worksheets relate strongly to the test material and format. In addition, students in lower-level courses often have difficulty learning from the text or classmates' notes, which may be less complete than those you provided in class. Even if you post your lecture notes online, these would not typically contain the class discussion that arose from questions you and the class discussed. In some instances, you may have a student that is gifted enough, or simply already knows the material well enough, to earn the desired grade without good attendance. Consider how you might feel if a student is able to perform satisfactorily on all graded papers for the class with scant attendance. Would you want to significantly lower the course grade?

A rigid policy which lowers the course grade based on a specific number of absences requires assessing the validity of absences throughout the semester. If you are considering doing so, you should check for any guidelines or requirements of your institution. Another option for having attendance impact the grade without such a rigid approach is to consider it part of your participation grade. Clearly an absent student is not participating in class discussion.

Some students lack the discipline to attend class if there is no immediate consequence for an absence. If attendance does not formally enter your course grade, consider keeping an especially watchful eye on first-year students. Your institution may have a formal reporting system designed to alert administrators and advisors to students who may be starting to falter in attendance and/or performance. These students often realize too late that they may not succeed in class without proper attendance. Even highly skilled students will often perform well below their ability when skipping class routinely.

Giving frequent quizzes, announced or unannounced, is one way to curb excessive absences without formally incorporating attendance in your course grade. This is especially true if the nature of your quizzes extends beyond rote exercises from the text. Incorporating points that arose in class discussion encourages attendance and attentiveness.

Taking attendance every day, even if it is not a formal component of the course grade, sends the message that you value attendance. It can also be valuable to know who has missed a quiz, handout, or an important announcement – or who did *not* miss one of these. You may be required by your institution to alert when students have not attended class, in which case this record keeping will be necessary.

You may also want to consider how tardiness might factor into participation or attendance. The distraction of a student entering class late can be a detriment to your concentration as well as that of students. Encourage students to alert you to problems that may cause frequent tardiness, such as a prior class reasonably far away from yours or one that frequently dismisses late. Even if tardiness plays no role in a student's grade, it may preserve the rapport you feel with a frequently late student when you know the cause.

You may be tempted to assume that since your students are adults, they will know what is expected of them. While some conduct should be obviously out of bounds, you owe it to your students to clearly state your expectations. It is also helpful to point out why you care about the guidelines you set forth. Generally speaking, our policies should create an improved learning environment for the class. If you are requesting students arrive on-time, explain that it is a distraction when a student enters late. In all cases, your tone should reflect that you are speaking to adults.

Pros and Cons of Grades Affected by Participation and Attendance

	Pros	Cons	Things to Consider
Participation	Encourages students to engage in the material May encourage students to attend office hours May encourage peer interaction when given the opportunity to work with peers in class	May be difficult for students who are uncomfortable speaking in front of the class May not properly credit students who are engaged but not as quick to offer immediate participation in class discussion	Students with disabilities may be at a disadvantage in regard to engaging in immediate classroom discussion, especially if they require additional time to process the material at hand. Consider offering a variety of opportunities for students to demonstrate engagement and participation, such as in-class activities, online/email communication, and office hours.
Attendance	Encourages routine attendance May provide better continuity for classroom discussion	Requires determining which reasons for absences merit being excused	There are a variety of valid reasons for missing class. A guideline must be in place at the start of the term to ensure that your decisions on excused absences are consistent.

Schedule

It is best to start the semester with a master calendar of planned coverage, homework or project due dates, and the dates of quizzes and tests. If you are teaching a coordinated class, much (or all) of this may have been already determined by the coordinator. If you need to decide any of the above, note that some may need to remain flexible, such as due dates on small assignments, dates of quizzes, and even the schedule of coverage. Others, such as exam dates or due dates for major projects, should remain fixed, whenever it is reasonable to do so. Students should be able to plan well for components that weigh heavily in the grade, especially as they balance their workloads between courses.

While it is important to try to stick to the schedule you lay out in your syllabus, cancellations due to weather or illness may derail your best efforts to do so. Despite the goal of keeping major dates fixed, you may want to offer a disclaimer that exam dates are tentative and clarify how you would announce any changes. For lesser components of the course grade, you may not need to give precise dates, provided you clearly state an intended schedule. If you plan to give quizzes each Wednesday, you may want to say, *"typically* we will have a quiz each Wednesday" or even just "we will typically have a weekly quiz." This gives your students a sense of the routine and allows you latitude to alter the schedule occasionally, without the class feeling misled or confused.

Similarly, while you need to outline the frequency of assignments, you may want to avoid exact dates for daily assignments. It may be difficult for you to determine exactly which problems you will want the class to submit each day or each week of the semester. Even a seasoned professor will need to occasionally alter a well-honed schedule due to an unplanned class cancellation or when an activity runs long. If you have not specified due dates in advance, you can announce "you should now be able to complete assignment #6, which will be due on Friday" or "you should now be able to complete assignment #6, through problem #24." Listing the assignments under headings for course subtopics will give students a general sense of how the assignments will be arranged, but gives you flexibility and may help you to avoid feeling rushed in class to adhere to a strict schedule.

Another option is to post daily homework assignments online shortly after each class meeting. This allows maximum flexibility to frequently adjust assignments. You can add a problem when an interesting example comes up in class, delete one if you did not have time to discuss a topic, and adjust the difficulty of your typical assignment for a particularly weak or strong group of students. This approach avoids having to announce changes made to your original plan, which eliminates confusion and the risk of students missing a change to a pre-printed assignment sheet.

If you are teaching more than one course, you may want to avoid simultaneous due dates. This serves the dual purpose of avoiding a heavy load of grading all at once, as well as spreading out the demands to assist students. One of the most valuable suggestions I received from student evaluations was to avoid giving tests in more than one course on the same day, so that fewer students would need to attend office hours at once. This practice has provided more focused and relaxed meetings with students preparing for exams and quizzes.

Pace is a challenging aspect of teaching for many and it impacts your ability to maintain an intended schedule. In Chapter 6, I will discuss a comprehensive practice of evaluating your teaching, including a course-end review. As you progress through the term, you will need to monitor your schedule, making adjustments where you have the freedom to do so. Making notes of problematic areas regarding your intended schedule, as they occur, may assist you in scheduling more accurately in the future.

Quick Glance: Tips on Scheduling

- Make a master calendar.
- Keep the dates of major components fixed, whenever reasonable.
- Keep the dates of lesser components flexible but provide a clear description of the routine.
- Consider announcing or posting specific assignments after each class period.
- If teaching more than one course, try to plan assessments so that grading is staggered, when possible.
- Routinely make notes on how scheduling could be improved and revisit your plan at the end of the semester.

Make-ups and Late Papers

Determining a make-up policy can be challenging. There are many instances in which allowing a make-up is appropriate but the process of weeding through the spectrum of excuses in a fair and consistent manner is not simple.

A graduate student with one course may find it relatively easy to offer make-up quizzes to all absent students, while a faculty member with three or four courses and a hundred or more students may find this daunting. How will you accommodate students who miss exams and quizzes or fail to turn assignments in on-time? Will the grade be a zero? Will you allow a make-up? If so, will it be for full credit? Will you drop a pre-determined number of scores to account for unavoidable absences? Whatever policy you set, make sure it is one you can apply fairly to all students and will be able to honor regardless of how hectic a given week may be.

In determining make-up policies, you should also be aware of any requirements at your institution. For instance, it may be expected that students who miss classes due to university-related events, such as away games for athletes, be permitted to complete make-ups provided they give appropriate notice. Start with a firm knowledge of what your institution requires and any general policies your department may follow, before asserting your own set of rules.

To avoid having to handle each incidence on a case-by-case basis, try to establish an across-the-board policy that you are comfortable following. On the first day of class, I acknowledge that there are valid reasons for missing a quiz, such as illness or family emergencies. I explain that instead of having students take make-up quizzes, I drop the two lowest scores from the semester for the entire class. This additionally assists students who were able to attend a quiz but had their preparation time affected by similar occurrences.

Having a clear policy in your syllabus is essential. As I discuss the above policy on the first day of class, I stress that I am not dropping quiz scores to ignore poor performances and the *sole* intent of these dropped scores is to account for unavoidable absences and impacted study periods. Everyone nods along as they are pleased to hear that they will get two of the lowest quizzes dropped and yes, that is all they hear of what I said. I receive several requests during the semester for the opportunity to make up a quiz because a student had a good reason for missing class and doesn't want to "waste" a drop quiz. I am able to refer students to the syllabus and remind them of the clearly stated policy, which includes the notion that the dropped scores are only intended to cover unavoidable absences and impacted study periods. Having this in writing can help assure a student that you are not making a judgement on the validity of their excuse but following a predetermined plan for such instances.

If you find a policy you have been following is problematic, seek guidance from colleagues or superiors as to how you might rectify it. If you plan to alter the policy mid-semester for one student, consider whether you are now obligated to reassess past decisions and whether you must announce a general change in policy to the class. Such mid-semester changes should benefit students and not impose anything stricter than indicated on the syllabus. A tightening of policy must wait until the next term.

> **Quick Glance: Tips on Determining Make-up or Late Paper Policies**
> - Learn the policies of your institution and department.
> - Be realistic about your ability to keep up with make-up quizzes or late assignments.
> - Determine a policy which can be consistently applied equally to all students. Avoid making exceptions on a case-by-case basis.
> - If you realize that a policy is detrimental to the class, seek a solution from colleagues/superiors and address the issue.

Calculators

Be specific as to the features students will need on their calculators, such as powers, log functions, trigonometric functions, exponentials, etc. Otherwise, your students may bring calculators that only do arithmetic and do not preserve order of operation or conversely, may bring in devices such as graphing calculators or plan to use apps on cell phones that provide more functions and detail than you may wish to allow. You may also want to discuss calculator policy with your colleagues. If you plan to allow graphing calculators but find others in your department do not, make sure to test that your students can work independently of these instruments as well. Otherwise, you may be sending your class off unprepared to future instructors who do not allow these devices. Conversely, if your department expects students to be educated on how to use graphing calculators in the context of the material in your course, then you will need to provide the proper instruction.

Some professors rely heavily on calculators while others disallow them altogether. Once in graduate school, I was teaching a remedial precalculus course and wanted to remove calculator use on an upcoming quiz or test. I cannot recall now why I wanted to do this on that particular paper, presumably because I wanted to test a specific ability to calculate something on their own. Students became concerned and whipped out their syllabi, which included a statement such as "students may use *non-graphing* calculators only." I meant the statement to disallow graphing calculators but I acquiesced that this could be read to assure the use of a calculator device. I have since added a statement to my syllabi that there may be instances when students are not allowed the use of a calculator but ironically have never felt the need to do so since. It was clear to me in the instance described above that, due to the remedial level of the class, the psychological stress that would be induced was not worth the element I wished to test and would likely have impacted students' overall performance.

Laptops and Tablets

If your course will be utilizing mathematical software, using online content, or accessing an e-text during class, allowing laptops in the classroom may be necessary. If not, you should give some consideration as to whether you will allow students to use a laptop or tablet for notetaking. There are at least two concerns to consider: the privacy of students with disabilities and the overall learning environment.

If you choose to ban laptops, you may still need to permit their use to students with certain disability accommodations. If a formal ban is in place, you may inadvertently reveal the presence of a disability for a student using one and privacy laws will leave you unable to directly address a perceived inconsistency. You could opt to allow laptops only on a case-by-case basis, requiring students who want permission to use a laptop in class to make a case to you in private that they are capable of taking effective math notes on the computer. This would lead to some ambiguity when students are seen using a laptop, rather than a clear indicator that the user has a disability, but could still result in challenges.

Students may be better off handwriting mathematics notes in class. Studies have shown decreased performance when notes were taken on a laptop versus long-hand (Mueller & Oppenheimer, 2014). Typing notes may require little concentration on the subject matter and students may transcribe rather than process and summarize the way a student writing out notes may. While long-hand notetaking may be more cognitively demanding, the resulting notes have been shown to be more beneficial for subsequent review (Luo et al., 2018; Mueller & Oppenheimer, 2014). This issue may not affect a student who takes notes on a tablet but the device is still potentially problematic.

Laptops and tablets provide the distraction of e-mails and texts to even the well-intended student who resists internet surfing. Many of us can relate to becoming sidetracked when such messages come in as we are desperately trying to get work done. Research has shown that multitasking on a laptop during a lecture impairs learning (Sana et al., 2013; Wood et al., 2012; Hembrooke & Gay, 2003). Instant messaging on a laptop during a traditional lecture style class can negatively impact performance, as may even course-related multitasking activities (Kraushaar & Novak, 2010). In comparison to students without access to technology during class, even students who used tablets flat on their desks (to allow observation of activity by the professor) have shown decreased performance similar to peers allowed unrestricted use of computers (Carter, Greenberg & Walker, 2016).

There is some evidence that not only the laptop user is affected. Students who are seated in view of another student's laptop have shown decreased learning as well (Sana et al., 2013). Classroom constraints may not afford every student who would like to use a laptop the opportunity to do so without affecting the learning of another student. If you do choose to disallow computer or tablet use in class, explain your reasoning. Some professors present this data to their classes and ask students to assist in designing a suitable and fair policy on the use of technology. For your first few courses, it is probably best to seek advice from colleagues and superiors on a policy. If you decide to allow for students' input in the future, your experience may indicate certain requirements are prudent.

Due to the aforementioned difficulty in typing up mathematics notes, you may not need to address the question of laptops directly in your syllabus, but you should think about these issues and how you might address them if they arise. It is likely that some students will indicate they might like to reference an e-book during class. Check in with other professors in your department. Discuss how frequently it is an issue and how they have dealt with it in the past. The more thought you put into such policy in advance the less likely you are to create unintended difficulties by answering a request off-the-cuff.

Phones and Smart Watches

Unlike the accommodation for laptops, you are on safer ground disallowing the use of phones and many smart devices. Phone use is often easy to note as students sometimes seem oblivious to the fact that they are not viewing us through a one-way mirror. Yes, we can see them, too! And yet, we are trying to teach, not police the use of devices. Since multitasking negatively affects learning and subsequent retrieval (Rosen, 2008), distractions, such as phones and laptops, may impair students' abilities to process and store the information they are trying to learn. This suggests restrictions on devices might assist our students' learning. Interestingly, at least one study found students with silenced, un-accessed phones may be just as distracted as those permitted to use phones. When researchers asked students to watch a 20-minute talk and take a subsequent quiz, performance was only improved for students who were not in possession of phones at all (Lee et al., 2017). In many settings, it is impractical and unrealistic to expect students to not carry a phone to class, nor would one typically require students to surrender phones to a teacher at the undergraduate level, so there may be no simple solution to this particular distraction.

If you want to use phones for activities in class but do not want them out otherwise, you may want to address this in your syllabus. Perhaps you want students to be able to use their phone for a calculator or to utilize a specific application, such as for polling student responses to questions posed in class (discussed in Chapter 7). If you want to use the phones throughout class, you will have little option but to simply allow the devices to be out. If you only want to do a few short exercises, you might place them at the start or end of class and have students put the phones away otherwise.

Smart watches can be a problem during testing, as students can view photos, receive texts, and access the internet. You should watch your class during a test to observe if students have questions and for security purposes. With small to moderate size classes, the inappropriate use of a watch should be fairly easy to detect.

Formulate a policy, which is in agreement with any held by your institution, and consider including it on your syllabus. Contemplate how you plan to address this policy. Will you actively speak to students as a class, individually, or possibly send an e-mail? Remember the intent of improved learning, not policing behavior, and maintain a positive tone as you address issues.

Conduct

It is important to discuss the expected behaviors in your classroom (MAA, 2018: 3) and you need to do so from the start. Your institution most likely has a policy on acceptable student conduct. You may want to expand on inclusivity and being supportive of other students during the class discussion. Adhering to your classroom policies falls under conduct, though violations may not represent unruly behavior.

You may feel that having policies regarding attendance, tardiness, and use of phones or laptops, fails to acknowledge that your students are adults, but these are put in place to improve the learning of the individual or class as a whole. Your students will soon be entering the workforce where there is a wide spectrum of expectations for conduct. Asking your students to adhere to a few guidelines, especially those backed by research to improve the learning environment, *is* treating them like adults. The crux is to have a legitimate educational or practical purpose behind your policies. Be transparent as to your motivation and avoid addressing these issues in a scolding manner. Remind students of policies and ask them *respectfully* to adhere to them.

Unexpected Issues

If students ask for your policy on a topic you have not yet pondered, consider telling them you will get back to them shortly, even if you have a gut response as to your answer. Giving yourself the opportunity to think through the implications of your decision may allow time to realize potential exceptions or to confer with colleagues. As long as you get back to students promptly with an answer you can abide by, this practice should serve you well.

> **Quick Glance: Course Policies Overview**
>
> - Learn your institution's requirements and restrictions before setting your own policies.
> - Determine what to grade and how to weight each. Consider the pros and cons discussed in this chapter.
> - Provide a clear but flexible framework for grades.
> - Determine a *realistic* policy for late or missed papers. You must be able to apply this policy consistently throughout the semester for all students.
> - Consider a wide variety of issues – even those which will not be formal policies included on your syllabus. Thinking about these in advance can help you determine whether they are worthy of a formal policy and shape your handling of issues that may arise.

Syllabi

In the introduction to this chapter, I stated that the syllabus is a contract with your students. This does not mean it must be entirely clinical. The statements in your syllabus are promises as to what you will teach your students, how and when you will assess their learning, and the policies you will follow for all students. These policies should assist in creating a good learning environment as well as avoid misunderstandings or confusion over the expectations in the course. You should also communicate how you hope the course will benefit each of them and how you want to help them learn the material.

Your institution likely has a set of guidelines and requirements for your syllabi as well as a supply of previous syllabi used for the course. If so, this is a great place to get started. You can begin by reviewing the contents of previous syllabi and drafting an outline for your syllabus which includes all of the mandated material. Reflect on syllabi you received as a student. Which were particularly illustrative and useful versus insufficiently informative? Elaborate on the required material, if needed, and add any additional topics which you feel are important and not yet discussed. If you have not been given any guidelines, provided below are some basics you should consider including.

If you have the complete control over course design and content, you may want to read through *Preparing for Class* in Chapter 3 prior to writing your syllabus. This book assumes you have little teaching experience and have therefore not been handed the task of designing an entire course on your own, but the concepts are similar to those discussed in planning an individual class. You will need to think of the learning objectives you have for your class, as introduced in Chapter 1, and apply considerable thought into how you will strive to achieve them.

Drafting a Syllabus

Now that you have thought about a whole host of issues that may affect your classroom, construct a draft. In each section, check that you are maintaining *clarity* for students and *flexibility* for yourself. At the top of the page, possibly in larger type or otherwise visually highlighted, provide the who, what, when, and where of the course and office hours:

- Course name and number
- Class meeting time and place
- Contact information (name, office location, office phone number, e-mail address)
- Office hours – indicate how changes or cancellations will be announced

Provide the general course information and required tools:

- Prerequisite course(s)
- Course topics: This could be your list of learning objectives or a more succinct, general list of the topics to be covered. Give sufficient information to express the material to be taught but avoid excessive detail if you will be articulating the learning objectives or desired learning outcomes elsewhere in the syllabus.
- Textbook: Specify title, author, edition, and publication year. Indicate whether alternate formats of the text are acceptable.
- Required materials (such as a calculator, including necessary/permissible features)

Provide the required components of the course, relevant policies, and how the course grade will be determined:

- Grades: List each assessment and how each is weighted. State how letter grades will be assigned. Include specific cutoffs, such as "A = 93–100%, A– = 90–92%..." or "a minimum grade of 90% will guarantee an A–, 80% a B–...".
- Important dates: State the dates of all major exams or projects. You may want to include a note about how you will handle notifying students if these dates should change for any reason.
- State the typical rate of quizzes and assignments or state the intended scheduling.
- Course policies for graded assessments:
  - Policies on missed/late quizzes and assignments
  - Policies on missed/late exams and major projects
  - Calculator policy and specific description of the type of calculator allowed and functions required, if not already discussed above
  - Honor violations. State any policy your institution may have and how you will deal with instances of cheating (which must be in agreement with any formal procedure that your institution has in place).

If you think there could be any confusion about what constitutes cheating, for instance, on graded homework problems, then designate a time to discuss this more fully before the first assignment is due, but after the first day of class. Using a handout with examples might be a useful demonstration (Felder & Brent, 2016: 57–8), especially if you have students attempt to decide whether cheating occurred in each example.

Discuss office hours and opportunities for help:

- Explain the purpose of office hours. Students sometimes worry that they are disturbing your work, so it is helpful to explicitly state that you are there to answer questions.
- Consider inviting students to stop by in the first week to introduce themselves.
- Mention any resources your institution may have for tutoring.
- Accommodations for students with documented disabilities: Indicate how students should pursue your institution's accommodations and provide relevant contact information. State any requirements regarding the deadline for requesting accommodations.
- Procedure for cancelled classes: Indicate how cancellations will be announced and how related announcements or instructions may be disseminated.

Clarify procedures:
State any official conduct policies, as well as your own expectations:

- State your institution's conduct policy, typically found in a student handbook.
- State any additional comments you may have regarding inclusivity and creating a supportive environment.
- If you have not already stated your institution's honor code, do so now.

Address the ultimate goals of the course by specifying desired learning outcomes. This could be done earlier in the syllabus as you describe the course content, after assessments, or as a wrap-up statement. Remember to leave yourself flexibility! If your list of desired learning outcomes is especially detailed, it may be more advisable to reduce the list to select, major outcomes.

> - Desired Learning Outcomes: State specific abilities students should have at the end of your course. Try to avoid using words like "understand" in favor of concrete action verbs. Consider how students can demonstrate the conceptual understanding you desire.

Finally, you might attach a daily or weekly schedule for the course or provide this as a supplementary handout.

> - Daily/Weekly topics: It may be best to keep this general, perhaps giving a topic for each week in the semester, rather than a day-to-day schedule, especially your first time or two with a course. Each group of students you work with will be unique and your daily schedule will likely have a little fluctuation.
> - Assignment schedule: As discussed in the previous section, if you are posting a complete list of assigned problem sets, you may want to avoid assigning each a specific date.
> - Include a clearly visible warning about the tentative nature of the schedule and how changes would be announced.

In general, as you write your syllabus, give yourself flexibility where possible, but also be clear and direct, so that there is little room for misunderstandings or arguments. Consider giving a syllabus quiz, discussed further in Chapter 3, to clarify your policies and encourage questions on any points of confusion.

Allowing for Choice

In *The Spark of Learning*, Sarah Cavanagh discusses how students' learning and motivation is affected by their sense of control over course activities and the degree to which they find activities meaningful. She suggests that one avenue to providing students control is by building choices into the syllabus, such as choices on assignments and in-class assessments (Cavanagh, 2016: 150–1). Choices could be as simple as choosing which problem(s) from a given list to submit or as complex as determining a project design and topic, given broad guidelines. It is important that choice offerings are balanced, so you will have a better perspective on how and where to offer choices after teaching a course through a couple of times. If you see opportunities in a class, make note of them for future use.

A Promising Syllabus

The goal of what Ken Bain refers to as a *promising syllabus* is removing the notion that the professor serves a dominating role. In *What the Best College Teachers Do*, he notes that the best teachers' syllabi tend to include three components. The professors discuss the opportunities that the course offers, explain the activities (rather than "requirements") which would be done to achieve the goals outlined, and how they and their students would evaluate the learning and adjust throughout the term (Bain, 2004: 74–5). This type of syllabus removes any sense of professor versus student. It sets the tone from the start that the professor is there to be a part of the learning process, rather than a distant judge of success or failure, but it does not lower the expectations or standards. Again, some perspective on your course is necessary to fully execute such a syllabus but you can possess this attitude from day one.

Sample Syllabus

> Math 110.01 – Calculus I – Fall 2021 – Prof. Kelwood
> MWF 10 – 10:50a in 304 Woolton Hall
>
> Dr. Ruth Kelwood   Office Hours*: MWF 11:30a – 1:30p
> Office: 231 Boykins Hall   TR: 2–4p
> Office Phone: x6347   and by appointment
> E-mail: r.kelwood@university.edu
>
> *Please note that in the event that office hours must be altered or cancelled, changes will be announced in advance whenever possible, in class or on the class webpage.

Why Take this Class?

Calculus requires attention to detail, analysis of problems, proficiency with procedures, and conceptual understanding resulting in appropriate applications of the tools learned. We will discuss real-world applications of this content but you can take something away from this course, even if you never use calculus again! We will emphasize the importance of understanding why a problem calls for a certain tool and substantiating solutions with detailed work, which demonstrates a comprehension of the nuances involved. The abilities to analyze and address issues and to provide supporting arguments with critical details are skills, which can transcend the course in a meaningful way to assist in a variety of non-mathematical jobs.

Prerequisites

Completion of Math 109 or placement into Math 110

Course Topics (as time permits)

Limits and Continuity: Tables of values, rules for limits, and continuous functions

Differentiation: Tangent lines, definition of the derivative, differentiation rules, and implicit differentiation

Applications: Extrema, increasing and decreasing functions, first derivative test, concavity, and asymptotes

If time permits, we may cover optimization and differentials in Chapter 3.

Textbook

Kelton, <u>Calculus</u>, second edition, Routledge, 2019. You may purchase or rent this text in whichever format is most convenient and economical, as long as it contains the complete contents of the text. Math 111 next semester will use this same text.

Calculators

You will need a **scientific calculator** for portions of the course material. Your calculator should be able to do powers (i.e. it should have a button like ^ or y^x), be able to take n^{th} roots (button like $\sqrt[x]{y}$ or $y^{1/x}$), and should include trigonometric, exponential, and logarithmic functions. Graphing calculators may not be used on any graded papers, since there will be many instances when you will be tested on skills, such as taking derivatives and graphing, that the calculator can perform. Please note:

- You are responsible for bringing an approved calculator to tests and quizzes.
- You may not share calculators during tests or quizzes.
- You may not use a graphing calculator.
- You may not use a phone, smart watch, or any other smart device as a calculator.

Grades

There will be weekly quizzes and assignments, two exams during the semester, and a cumulative final exam. Your course grade will be based on the following breakdown.

Homework average	20%
Quiz average	20%
Two exams each 20%	40%
Final exam	20%

Letter grades will be assigned as follows. A: 93–100%, A–: 90–92%, B+: 87–89%, B: 83–86%, B–: 80–82%, and so on with <60% being recorded as an F.

Office Hours: Come on in!

The office hours stated at the top of this syllabus are times that I have specifically set aside to answer your questions on class discussion and homework problems. You can come alone or with other classmates and you don't need to make an appointment for the times posted on the syllabus. I am just sitting there waiting for your questions! Please consider stopping by some time in the next couple of weeks to ask a question or just to introduce yourself. If you have conflicts with the times offered, please request an appointment.

Academic Support Center: Free Peer Tutoring Available!

Free tutoring is available by appointment in the Academic Support Center. Please consider using this resource in addition to attending office hours.

Participation, Attendance, and Tardiness

It is necessary for your success in this course that you be engaged during class. For this to be possible, you must be present! <u>Attendance is expected for every class meeting and students are expected to arrive to class on time to minimize distractions and interruptions in our class discussion.</u> You do not have to know the correct answer to participate! Ask questions, discuss problems with classmates during in-class exercises, and offer *possible* ways to start working on a problem.

Homework

Homework will be assigned after most classes. The assignment will be available online after 11:30a on the day of class on our class page in your university portal account. Each Wednesday, the assignments from the previous week will be collected at the start of class.

In addition to attending office hours, you are encouraged to work with other students when studying and working on homework concepts, including seeking help from a tutor at the Academic Support Center. Any work submitted must be your own and you must therefore understand and be able to explain any assignment turned in for a grade. Any work submitted which is not your own constitutes an honor offense.

Late Homework

Late assignments will not be accepted. To account for unavoidable absences such as **illness, family emergencies, etc.**, I will drop the two lowest homework scores for each student. These dropped grades are not allotted to ignore low scores, but rather to prevent unavoidable absences/circumstances from lowering your grade.

Quizzes

Typically, we will have a **quiz each week**, based mostly on the material covered since the last quiz/exam but may also contain material from earlier in the term. We will have *approximately* 12 quizzes. Currently, most quizzes are scheduled for a Friday but class cancellations and other scheduling issues may alter that schedule. *****Students with documented disabilities must make arrangements with me prior to the start of a quiz or test.*****

Missed Quizzes

You must be present to take a quiz. No make-up quizzes will be given, except in the case of missing class due to an approved event sponsored by the university. In this case, you must have given me prior knowledge and discussed plans for a make-up *prior* to your absence at least 48 hours in advance of the absence. To account for unavoidable absences such as **illness, family emergencies, etc.**, I will drop the lowest quiz score for each student. This dropped grade is not allotted to ignore a low score, but rather to prevent unavoidable absences/circumstances from lowering your grade.

Exams

The tentative exam dates are October 6, November 19, and finals week. The final exam is cumulative. If I need to move an exam date, I will give as much notice as possible before the new test date. Please see the next section regarding missed exams. *****Students with documented disabilities must make arrangements with me prior to the start of a quiz or test.*****

Missed Exams

If you miss an exam, you are not guaranteed a make-up exam. Prior or immediate notice must be given for a make-up to be considered. Make-up exams are subject to point penalties and/or may contain different content. Failure to notify me within 24 hours after the missed test may forfeit the opportunity to take a make-up exam.

Students with Documented Disabilities

If you have a disability which requires accommodations, please contact the Undergraduate Studies Office (x5555) to establish the specific adjustments necessary. If you wish to utilize these accommodations, you must personally notify me *prior* to the test or quiz in question and make the necessary arrangements at least 48 hours *in advance*. Once you have started work on a test or quiz, you must complete your work at one sitting. You may not resume work at a later time. I want you to have the time you need, so please schedule in advance!

Class Cancellations

If class is cancelled for any reason, we will typically resume the course as planned for the day cancelled on the next day we meet, unless announced otherwise on our class webpage or in a previous class. If class is cancelled, please check the webpage after 10a for any announcements.

Announcements

Announcements will be posted on our class webpage, so please check it regularly. If you miss a class, it is your responsibility to contact me or a classmate about any possible in-class announcements that you may have missed.

Conduct in the Classroom

Professional conduct, including respect and consideration for all those in the classroom, is expected. In an effort to provide the best learning environment, it is requested that you arrive on time and put away all electronic devices. If you have concerns about the classroom requirements or environment, please speak to me privately as soon as possible.

> The University Civility Policy:
> ~ *Quoted institution's policy here* ~

Honor Code

Any student who participates in a violation of the University Honor Code will receive a zero on the paper in question. This includes students who give unauthorized aid as well as those that receive it. No aid is permitted on any quiz or exam, including the use of any device other than a non-graphing calculator to complete problems. Phones and other smart devices are not permitted to be used as calculators. Homework submitted must represent your own work.

A zero received due to an honor code infraction may not be dropped

> The University Honor Code:
> ~ Quoted institution's policy here ~

Learning Outcomes for Math 110 Calculus I:

Students should be able to describe the notion of a limit and provide a real-life example. They should be able to evaluate limits using limit rules, the Replacement Theorem, graphs, and numerical techniques. They should be able to state the formal definition of a continuous function and its meaning in relation to the graph. Students should be able to determine the continuity of functions, using the formal definition, the function definition, and the graph.

Students should be able to state the definition of the derivative of a function and be able to find derivatives for polynomial, rational functions, and radical functions using the definition. They should be able to state the meaning of the derivative in terms of tangent lines and rates of change.

Students should be able to demonstrate the techniques of differentiation, including the power rule, product rule, quotient rule, and chain rule. They should be able to use these techniques to find the derivatives of a variety of algebraic functions, including polynomials, trigonometric functions, exponential functions, logarithmic functions, and algebraic combinations of these.

Students should be able to use derivatives to find extrema and inflection points and be able to describe the function's behavior and contour on intervals. They should be able to use the information gained from analysis of the derivatives to construct the graph of a function.

Catalog Description: MAT 110 Calculus I

> ~Insert institution's description here~

The Basics of the Classroom   3

The First Day of Class

I recall my first day as a graduate teaching assistant running a calculus discussion hour as exhilarating and slightly terrifying. At the time, I held many uncertainties about my future. My ability to successfully complete my graduate course work, perform mathematical research and write a dissertation, and succeed as a professor was unknown. Entering the classroom that day was the beginning of my career – I hoped. Now, over 20 years later, I can say that my first day of class is still filled with nerves. My stomach churns with unease every time I meet a new group of students. Why, when I am an established and tenured professor, would this still occur?

I had assumed for some time that these persisting first-day nerves were due to my general tendency towards anxiety until one two-minute conversation completely altered my outlook. Several years ago, I was walking to class with a colleague of mine who had been teaching for decades and who carries a strikingly calm and centered air. I mentioned my inability to shake my discomfort on the first day of class in hopes that he would offer some sage advice. To my great shock, he said that he thought *anyone who was not nervous on the first day probably did not belong in the classroom anymore*.

I took three notions away from this brief exchange. The first was the realization that he, too, experienced this nervousness after so many years of teaching despite being a very relaxed and confident professor. The second was that I am normal, which is always a comforting thought, and the reaction was not demonstrative of a deficiency on my part. Finally, it led me to a conclusion about why nerves could be indicative of a better professor and a lack thereof might suggest a waning career.

On the first day, I am meeting a group of people with whom I will work for the next several months. Their level of enthusiasm and willingness to engage will affect my experience as much as my skill and demeanor will affect theirs. I want my class to see right from the start that the course will be efficiently run, material will be presented accurately, and questions will be met enthusiastically. It can be challenging after an extended break between terms, to present a smooth and effortless flow to the disjointed nature of a first class in which I am taking attendance (stumbling over pronunciations), handing out and reviewing syllabi, and engaging students in a brief lesson on mathematics. Despite my strictness with the course policies and the high bar I hope to hold for the mastery of skills in my course, I want my class to understand how much I desire their success. Thus, not only am I peeking at how enjoyable I might find this particular group, but I am also feeling the pressure to showcase the best of what I hope to offer in the coming months. The lack of *any* nervous energy might signify a high level of confidence, but it may also evidence a disinterest in the student experience.

Getting Class Started

Arrive early. This not only affords you the opportunity to get organized, but gives students a chance to touch base if they are attempting to add into your class or have questions about whether they should enroll. (Plus, racing into the room at the last moment, out of breath, may not make the best first impression). Write your name and the course on the board before class so that students can confirm they are in the right place when they enter the room. As you await the start of class, you may be absorbed in thoughts about all you want to say and do in this first class. Be mindful that students are already trying to assess what this experience will mean for them. Consider the different initial impressions of a professor smiling at students as they enter versus one tensely reviewing a checklist.

Start class with a brief period of introduction, review your syllabus, and get to a little math! Share a little about yourself and, depending on class size, you might ask students to introduce themselves to you. You will have a lot of information that you want to convey to students as you go over your syllabus. Be mindful that they may be attending four to six courses in a couple of days' time and will be receiving a barrage of information. You may find there is not too much time for mathematics on the first day, but it is important to work some lecture or exercise into the first class.

Getting to Know Each Other

Telling students where you are from, why and how you were led to teaching mathematics, where you may have attended college or graduate school or your history at the current institution, and your general interests is a nice way to personalize yourself right away. Keep it somewhat brief on day one, as you will have a lot to do this class period.

Take attendance. If your class is small, you can call out names from your roster or have students introduce themselves to you. For larger classes, you could pass around a sign-in sheet. If your class enrollment is sufficiently low, say ten students or less, you could have each student also share a personal interest or intended major. An alternative is to have students fill out a note card with this information, send it in an e-mail, or submit it (privately) through an online course page. These options provide a written record that you can review after class, work in larger classes, as well, and may encourage reserved students to share more freely. Learning a little about your students may inform you as to application examples they might find intriguing and conveys your interest in them as individuals.

You can also use note cards (or the above-mentioned alternatives) to learn background information or potential concerns. Students can list their last math class and how long ago it was. You could ask them to indicate a letter grade which they feel reflects their *comprehension* of that material which may differ from the actual grade received. This eliminates the need a student may feel to defend a poor performance or explain that a high letter grade was not representative of his or her understanding. Students may also use this as an opportunity to alert you to demands they might face outside of class, such as employment, participation on an athletic team, or family responsibilities.

You can extend this exercise to the expectations and concerns students may have as they enter your course. In *Establishing a Comfortable Classroom from Day One: Student Perceptions of the Reciprocal Interview*, the authors describe an exercise in which students are asked to state their goals and reservations, indicate how the instructor can help students achieve goals and what resources they bring to the class, specify what behavioral norms should be established, and discuss any dislikes they have of professors and classes (Case et al., 2008: 211). The authors describe this as a 30-minute exercise in which students work individually and in groups, then report the group findings to the class.

Another 20 minutes is used for students to formulate questions and interview the professor. The time frame could be significantly compacted by having students submit these answers on note cards or online, with the professor offering a summary of student comments and answering selected questions in the next class meeting. If you wish to preserve the social gains of the exercise, maintaining some group discussion and addressing at least some portion of the comments or questions in class may be best.

Syllabus Review and Setting the Tone

It serves little purpose to read the syllabus to students *verbatim*, but there is value in going through each section of your syllabus, stating the main points, and summarizing what you have carefully written out on your syllabus. Not only can you confirm that you have stated these policies directly to your students, but students have an opportunity to question anything which seems immediately unclear.

If you included a conduct policy in your syllabus, this provides an opportunity to discuss the social expectations you have for the classroom. You can explain the importance of attendance and punctuality, but more importantly, you can assert from the first day the need to be respectful of other students' thoughts and questions. Doing this on the first day of class demonstrates that this is a priority and may have greater impact setting the overall tone for the class. In *How Learning Works*, the authors note:

> First impressions are incredibly important because they can be long-lasting. Your students will form impressions about you and the course on the first day, so set the tone that you want to permeate the semester
> (Ambrose et al., 2010: 184)

It is a great time to inject your desire to hear their input and address any confusion on material. I often tell students that I have never heard a "stupid" question. While I may sometimes be surprised a student does not know some fact, my interest is purely to clarify or correct the misperception. Unexpected questions are informative; they instruct me as to what concepts may not have been covered sufficiently or background information students may be lacking. If you did not include a conduct policy in your syllabus, take time at some point before you begin the mathematics discussion to express your expectations for the classroom environment. This is also a good time to try to ease anxieties, especially in entry-level mathematics courses. Promote a "growth mindset," the notion that students can improve their skills with hard work and proper study skills (see *Enthusiasm & Motivation* later in this chapter).

Often students will not ask many questions during the initial syllabus review, so it is useful to follow up with a "quiz" of some type. This might be simply a few quick questions posed to everyone at the start of the next few class meetings, such as "do I accept late assignments?," "can you make up a missed quiz?," or other general policy issues. Alternatively, you could have a more formal, graded quiz, intended to be an early "gimme," which could be open-syllabus. The point is to have students familiarizing themselves with the contents of the syllabus and your policies after the initial wave of information from all of their courses. This review offers them a greater chance of connecting the information with your class. It is an excellent opportunity to not only review, but to encourage questions and discussion, if students answer incorrectly. This is always best when the issue is merely theoretical and before it is directly impacting a student who is making a request counter to the policies stated in the syllabus.

Teaching

Give at least a sample of what your course will feel like. Students usually have only a few class meetings to decide if a course is a good fit for them and subsequently enroll in a different course if it is not. They will be assessing the level of the material and your particular teaching style, so be clear when something is intended as review of prerequisite knowledge versus new information. You likely will not have a significant amount of time to teach on the first day, so choose your material thoughtfully. Select a lecture topic and/or in-class activity that can be completed in the short time you have for content. If you want students to play an active role in class discussion or content discovery, then choose an activity which will encourage this on day one. Foster the supportive atmosphere you indicated on your syllabus that you expect for your classroom, by soliciting responses and responding positively. Consider how you might wrap the class discussion by posing questions which will be addressed in the next class or the homework assignment.

Prerequisite Quiz

Your class may be composed of students with varying mathematical backgrounds. Administering a quiz on prerequisite materials can be eye-opening for you, especially if a significant portion of the class is underprepared, and may alert students who lack the appropriate foundational skills. An instructor must not only assess the amount of knowledge, such as the breadth of procedural knowledge, but also the nature of that knowledge (Ambrose et al., 2010: 20). Are they proficient with procedures and do they understand the related concepts/applications, or have they only learned the material on a superficial level?

You may not want to start the first day with a quiz. There is a lot to do that first day and receiving this information in the first week should be adequate. You can make this a homework assignment if you would like to lessen the stress students may feel, but explain to students that you want their best attempt without assistance from others because you will be using it as a guide for future lessons.

If students were not given an opportunity to prepare for the quiz, a grade based on accuracy should not count towards the course grade unless the class is subsequently able to earn full credit through corrections or a re-quiz. If you are troubled by how students performed on the prerequisite skills, then discuss the common errors in class and consider giving an announced, graded quiz on the material in the near future. Clearly state which skills will be tested on a review handout and offer to assist anyone in office hours who needs instruction (Felder & Brent, 2016: 61).

In Chapter 1, I discussed how drawing connections and parallels between previously learned material and the new concepts at hand can improve learning and achieve *Integration* and *Caring* in Dee Fink's Taxonomy of Significant Learning. Knowing how well your students learned and remember this earlier material is essential in trying to draw these connections in an effective way.

> If [students] do not draw on relevant prior knowledge — in other words, if that knowledge is *inactive* — it may not facilitate the integration of new knowledge. Moreover, if students' prior knowledge is *insufficient* for a task or learning situation, it may fail to support new knowledge, whereas if it is *inappropriate* for the context or *inaccurate*, it may actively distort or impede new learning.
>
> (Ambrose et al., 2010: 13–14)

When you know in advance material which you would like to draw on in a course, it may be helpful to include it in a prerequisite quiz, even if the knowledge is not strictly required for the present course. It may give you a heads up to the need to review this earlier topic before attempting to connect it to the new material. Examples of connecting topics can be found in Chapter 5.

How Did It Go?

I cannot say that I can recall any disastrous first days in the classroom, but some are less satisfying than others. The shiest classes can be the most challenging and they can make for an awkward first day. At the start of your career you may find it difficult to accurately assess why the vibe may be off. Try to note moments that were highs for you and consider what lead to them. Similarly, could you have avoided any of the lows that may have occurred?

Quick Glance: First-Day Overview

- Arrive early.
- Welcome students and give a brief personal statement.
- Take attendance in a manner appropriate for the class size.
- Ask students to provide information on their math background, outside demands (job, athletic team, commuter), and any personal interests that they would care to share. This could be on a note card in class or after class through email or a private online submission.
- Review the syllabus.
- Discuss the expected behaviors for the classroom.
- Give at least a brief lesson.
- If you want the class to be active and engaged, your first lesson should embody this!
- Consider a prerequisite "quiz" in the first week. This could be done online or as homework.
- After class, reflect on what went well and what may have been lacking.

Preparing for Class

Like many graduate students, my first teaching assignment was a discussion hour in which I answered homework questions for another instructor's calculus course. As I described at the start of this chapter, I was quite nervous. To temper and control that tension, I needed to feel prepared. I completed the students' assignment, which ensured that I was alerted to any formulas or clever approaches that might be needed before I was standing in front of the class. When I am preparing a lesson today, I still often start with the problem set available. It reminds me of the various nuances of the material I want to address in class and alerts me to any insufficiencies in the problem set. This process may also help avoid what is called "expert blind spot," in which instructors' sophistication with the material causes them to inadvertently skip over steps or details necessary for a novice.

Desired Learning Outcomes

As you prepare an individual lecture or classroom discussion, identify the desired learning outcomes. Start by creating a list of the skills and concepts you want students to learn (i.e. learning objectives), then create the list of desired learning outcomes by specifying how students will demonstrate mastery of these skills and concepts. List any issues within the content of which students need to be aware. If you are having difficulty getting started, you may find it helpful to turn to the homework assignment, as I described above.

If the assignment has been pre-selected by your department, reviewing these problems will give you a sense of where the department wants you to focus and the specific skills it expects your students to master. If you are selecting problems yourself, determining which problems appeal to you in the available problem set may guide your lesson, stimulate ideas for in-class problems, or make you aware of a deficit in the text's coverage. This process may impact the list of learning outcomes you generate, and it familiarizes you with how the text is approaching the material, as well as the terminology, notation, and instructions your students will see in their assignment. Whether or not you use the assignment as a launching pad, you will want to check the compatibility of your desired learning outcomes with the assignment prior to fully designing your lesson.

In Chapter 1, we considered possible desired learning outcomes for a calculus course which included the statement that students should be able to:

> - Find the limit of a function as the input approaches a finite value using a table of values, limit laws, the Replacement Theorem, and by inspection of its graph.

As we consider creating the specific lesson plans for this topic, we might recognize the need to specify additional outcomes.

> Students should be able to:
>
> - State the *y*-value on the graph of function, if one exists, for a given *x*-value, including where the function has a discontinuity.
> - Using a graph, find the limit of the function when approaching *x*-values, including where the function has a discontinuity.
> - Use limit laws to determine the limit of a function, if it exists.
> - Determine the limit of a function when approaching a given *x*-value, if it exists, using a table of values.
> - Know when and how to use the Replacement Theorem to determine the limit of a function.

These itemized outcomes expose various settings and conditions that we need to explore with the class. The order above avoids starting the class with the plug-and-chug table in favor of the graphical representation and a discussion of the concept of limit. Another approach could reorder the list entirely.

In the following sections, we will examine how we can build upon these desired outcomes to help craft the lessons for this topic. Design your discussion, activities, and homework problems to illustrate the concepts and practice skills (Felder & Brent, 2016: 20). Consider the balance of lecture and in-class activities you want to achieve and how your students will engage in the material.

Plans for Assessment

Prior to drafting a lesson, think about how you plan to assess the desired learning outcomes. In the context of the limit example, how will you assess students' mastery of the limit concept? Will you focus on tables, algebraic solutions, or analysis of graphs? Most calculus courses include all of these approaches, but the weight varies. While I teach my classes how to find a limit with a table and may ask them on a low-stakes assessment to demonstrate this skill, I rarely include it on an examination. My focus is on the algebraic and graphical computation of limits. This is reflected in how I approach the material in class. After considering assessment, review your list of desired learning objectives for completeness and accuracy.

Creating a Lesson

Having determined desired learning outcomes, you may find yourself drawn to writing out a lecture which takes a natural progression through the topic at hand, laying out the fundamentals, then building up content through a series of examples. This process may organize your thoughts and promote the discovery of the building blocks you need to use, but I encourage you to evaluate the student experience.

Consider the following preliminary plan that could be constructed based on the learning outcomes listed in the previous section.

Rough Outline of Building Limit Concepts

- Analyze graphs

Draw the graph of function which has a non-removable discontinuity and several removable discontinuities. Discuss the function value (or lack thereof) for various values of x, including discontinuities.

- Define the limit of a function.

Find the limit of the graphed function for each x-value discussed above.

- Discuss computing limits using a table of values.

State a rational function with a removable discontinuity and for which the Replacement Theorem will apply. Create a chart of function values for a short list of x-values before and after the x-value for which the function is discontinuous. Determine the limits.

- Discuss computing limits algebraically.

State limit laws. Work a variety of examples which utilize the limit laws.

State the Replacement Theorem. Compute the limit for the rational function example used earlier and compare to the previous method (using the graph).

- Return to graphical approach.

> Provide the sketch of the rational function and use this to compute the same limit.
>
> • State summary and conclusions.
>
> Review the different methods available to compute limits. Discuss when we might prefer to use one method over another.

While the above list outlines the path by which we can build a discussion of the concepts involved, it has not yet incorporated student engagement or interest – it does not even acknowledge there are students in the room! As such, this outline is lacking necessary components for quality learning. In *What the Best College Teachers Do*, Ken Bain describes the following framework in which people learn most effectively:

> (1) [T]hey are trying to solve problems… that they find intriguing, beautiful, or important; (2) they are able to do so in a challenging yet supportive environment in which they can feel a sense of control over their own education; (3) they can work collaboratively with other learners to grapple with problems; (4) they believe their work will be considered fairly and honestly; and (5) they can try, fail, and receive feedback from expert learners in advance of and separate from any judgment of their efforts.
> (2004: 109)

As you design a lesson, consider whether your students would experience this environment in your class and work to create as many of these experiences as possible. In the next section, *During Class and Office Hours*, I will elaborate on the classroom environment and the overall tone you set for your course.

If you find yourself developing a pure lecture, there are two basic steps which can allow you to move towards a more interactive class. First, determine points when you will be able to ask questions. Include *open-ended* and *divergent* questions (Cashin, 1995a). Open-ended questions allow students to answer in any form and are constructed as to avoid a simple yes or no answer. Divergent questions may have a variety of possible answers and seek insight. Second, consider where you might ask students to try an exercise on their own or in groups. When students are given opportunities to collaborate with each other in the classroom, they can test ideas and hear alternative approaches, which research has shown can improve conceptual learning

(National Research Council, 2003: 22). Can students work together to try to construct the solution to the motivating problem? You may have to fight the urge to tell your students all of the content directly, but the more interactive you can make the class discussion the more enjoyable and productive the class will be for everyone. A lesson with a high energy level and students engaged in learning promotes a powerful learning experience (Fink, 2003: 6–7).

Look at the dramatic difference we can create when we modify the rough outline to include students engaging in the material and proposing solutions:

- Analyze graphs

Draw a continuous graph and ask the class to state y-values on a graph of a continuous function for a given x-value.

Draw the graph of function which has a non-removable discontinuity and several removable discontinuities. Have students work in small groups to determine the y-value if one exists, for each x-value of interest. Include the discontinuities, as well as at least one x-value for which the function is continuous. Call for and discuss answers. (Alternatively, the skills here could be tested on an opening "quiz." Opening exercises will be discussed in the next section in *A Typical Class*).

- Define the limit of a function.

Have students return to the graph and work in small groups or pairs to propose the limit of the function as they approach each x-value investigated before. Call for and discuss answers.

- Discuss computing limits using a table of values.

State a rational function with a removable discontinuity. (Select one which reduces to a simple polynomial, but do not simplify it yet).

Ask students to create a chart of function values for a short list of x-values before and after the x-value for which the function is discontinuous. Ask the class to propose a limit based on their calculations.

- Discuss computing limits algebraically.

> State limit laws. Ask students to work together on a variety of examples which utilize the limit laws.
> State the replacement theorem. Ask students to assist in simplifying the rational function in the last example.
> Have students use their simplified expression to compute the limit and compare to their previous result. Call for and discuss answers.
>
> - Return to graphical approach.
>
> Provide a sketch of the rational function, using the students' simplified expression or ask students to sketch.
> Ask students to use the graph to compute the limit and compare to their previous answer.
>
> - State summary and conclusions (done at the end of each class, if covered over more than one class period, and after all lessons have been completed).
>
> Ask students to list the different methods available to compute limits.
> Ask students to suggest the settings in which they might prefer to use one method over another.
> Ask students when a table of values might be necessary.
> Create summary lists on the board as students answer the above questions or use for a closing "quiz." (Closing exercises will be discussed in the next section in *A Typical Class*).

Another way to create interest is by creating connections. In selecting the functions, consider an application which relates to majors in your class, if possible. This may be straightforward if you are teaching a calculus course designated for business, science, or math majors, but many lower-level courses may include a wide variety of intended majors and students with no clear career paths yet. Consider whether there are any parallels in your current topic to those taught earlier in the term or earlier in your students' math education.

Connecting new concepts to relevant material which has been previously learned (correctly) can not only help your students' learning, but their retention of this knowledge (Ambrose et al., 2010: 15). For instance, suppose you are teaching a multivariable calculus course and plan to discuss the second partials test. There is a clear relationship to the second derivative test from single variable calculus. If students understood that test well in the past, there is little adjustment needed in this new setting, accelerating their comprehension and recall of the second partials test. Recall, in the section *Prerequisite Quiz*, I mentioned the importance of assessing your students' background when trying to build upon or draw parallels to lower-level material. Some students may not have learned the second derivative test in their first-year calculus course. Others will have forgotten this test or may have had difficulty using it properly in the past. Typically, students in a multivariable calculus course should be capable of quickly comprehending the second derivative test and teaching/reviewing it may be a valuable steppingstone to the second partials test. Backing up a step to a simpler concept and then returning to the more complex setting provides a conceptual anchor. It transforms new information into something partially familiar and gives it meaning. Without meaning, without relationships, students turn towards memorization (Steele & Arth, 1998). You should not expect that students will draw these parallels on their own; it is important to facilitate these connections (Ambrose et al., 2010: 32). This does not mean you have to explicitly state the parallel, but help your students discover it. For instance, you could discuss the second derivative test and ask your students to make conjectures about the multivariable setting.

Another important component of your plan is to properly motivate the discussion. This might be with a new question or problem or you might revisit an earlier problem. This approach strives to engage your students and prime them to better receive and *process* the information you want them to learn. If students have offered a conjecture about a question, based on their previous knowledge, they are better prepared to understand the answer (MAA, 2018: 25). Can you craft a problem your students have enough knowledge to approach, but not solve without the upcoming topic? This is one way in which you can help students see the purpose of what they are learning.

> Most students, even those who desire to succeed in school, are intellectually aimless in mathematics classes because often they do not realize an intellectual need for what we intend to teach them.
>
> (Harel, 2013: 119)

Guershon Harel, a mathematics professor whose research lead to his formulation of a theoretical framework called DNR-based instruction, asserts that intellectual need is best achieved when a student engages in a problem which leads to the construction of the desired knowledge and an understanding of how the problem has been resolved (2013: 122).

Returning to the example of the calculus class, you might consider a broad question which motivates the concept of limit. A simple example would be to tell a class that a car is traveling along a highway with a speed limit of 65 mph when it notices a police car ahead monitoring speed. You can ask what they expect the car's speed to approach as it nears the police car, assuming it is traveling at least 65 mph. (Subsequently, you might inquire why the assumption was necessary. Would a slow-moving vehicle speed up?) You can follow with other scenarios, such as the speed approached when the driver nears a stop sign at the end of an exit ramp.

Discussing the notion of a limit in a clear everyday example has the power to drive home the comprehension of the core concept, when it is likely that your students have encountered these scenarios. When students grapple with why the function value need not agree with the limit, you can remind them that the car did not have to stop at the stop sign just because its speed approached zero. It is not what happens at the stop sign, but as you are rolling up to it! (It's not a perfect example in this sense since the speed of the car is continuous – and you should be willing to freely admit this or even ask them to find the flaw – but students often find this example illuminating). Consider reviewing the examples and activities planned to see where you can proactively work to pique interest and curiosity. For example, here is how one point above could be reworked:

- Define the limit of a function:
    - Describe a car on the highway and ask students to predict the value the speed approaches as the car approaches different points on the road.
    - Give the mathematical definition of a limit.
    - Have students write the answers for the highway example symbolically, letting the speed of the car be denoted by a function.
    - Sketch a graph that might represent this speed function and again refer to previous answers and how they would be reflected on the graph.
    - Have students return to the graph used for the opening exercise. Have students work in small groups to propose the limit of the function as they approach each x-value investigated before. Call for and discuss answers.

Injecting this simple example may make it easier for your students to approach the more abstract examples which may follow.

You may be asking, shouldn't we have started our planning with the motivating question rather than considering it after all of this work? It is certainly helpful if such an exercise comes to mind as you begin to lay out your ideas. The more experienced you become, the easier it will be to recognize on the outset how you might motivate a discussion. If you have not thought much about a topic in several years, you may need to dive back in and reacquaint yourself with its nuances. Revisiting old concepts often allows you to see a new dimension of them. As such, the motivating question might only emerge after spending time developing ideas on how to scaffold the necessary skills and concepts.

Finally, address the issue of time. If the class discussion takes off and you find there is less time than anticipated, what elements could be altered, eliminated, or assigned for homework? If the class moves more quickly than you had planned, how could you extend the planned activities or what additional exercises could you add?

Preparation Time

My husband is an experienced mathematics professor and has a 75-minute commute to his campus. Frequently, he is able to sufficiently design his lectures in his head as he drives to work. I need a pencil and paper to organize my thoughts and prefer to have completely written out examples in front of me for my reference during class. Regardless of the style of preparation that works for you, be well-prepared, but not over-prepared.

Robert Boice studied the careers of new faculty and discovered that those who spent an excessive time on course preparation injected too much course content. Their students were less engaged and ultimately less satisfied with the course experience (Felder & Brent, 2016: 43–4). Boice noted that the best new faculty he observed spent less than 2 hours on prep per hour of class time (2000: 11–12). Once you have taught the material repeatedly, you may not need nearly this much time to prepare a lesson unless you decide you want to tweak your approach on a topic.

One reason the excessive preparation is likely unsuccessful is that it may not factor in sufficient opportunities for student response. It does not take a significant amount of time to create a few examples that students might need five to ten minutes of class time to consider, though it may take some experience before you choose effective exercises that hit the points you want to cover efficiently.

To save time or gain inspiration, you might turn to resources that have already been tested and refined. Your text likely has supplemental resources and there are many online options, such as MIT Open Course Ware, Khan Academy, Google, and materials posted by a variety of universities. Your colleagues may have discovered or developed handouts they find useful. Your search may unearth the perfect resource or simply stimulate ideas of your own. Inject yourself into materials that you find, modifying them based on what you want students to take away from the discussion. Keep your initial list of learning objectives in mind! (Felder & Brent, 2016: 44).

Quick Glance: Preparation Overview

- List the learning objectives and desired learning outcomes for the topic to be covered. This list may need to be broken down over several class meetings.
- Examine the homework assignment if one has been predetermined. If you are creating the assignment, determine which problems you would like to assign.
- Review the list of desired learning outcomes for compatibility with the planned assignment and adjust, as necessary.
- Determine the strategy of exposition for the material.
- Review the lesson plan for opportunities for student engagement and discovery.
- Consider what connections to previous material or life situations may exist.
- Consider the student experience and how it might be more active and enjoyable.
- Devise a plan for adjusting for having more or less time than anticipated.
- Be prepared, but avoid over-preparing in an attempt to manage every moment of class – create opportunities for student engagement.

During Class and Office Hours

You have finally made it to the classroom with your well-prepared lesson in hand. Be ready to hand over some control to the students before you. *Steer the ship, rather than sail it alone*, and actively work to create an environment where students feel safe in offering conjectures. Turning to your students for input, not only allows them the opportunity to engage in the material, but conveys that you are interested in whether they are comprehending the discussion or have alternative approaches. You may find it challenging to avoid simply lecturing to your students initially, but the more you actively pursue breaking up your "speech" with engaging activities and student contributions, the easier and more natural this becomes. You will see improved attentiveness, participation, and learning, and you will experience a much more satisfying environment for both you and your students.

A Typical Class

The start of any class should invite and prompt questions. Some classes will include students who eagerly inquire about homework problems nearly every class meeting, while others lack more than one or two students willing to do so. Offering engaging approaches to reviewing the previous day's content or assignment will improve your students' ability to receive and process any corrections or clarification they need and may serve to prime them for the new material to be covered. Similarly, motivating the upcoming lecture by a thought-provoking question can capture students' attention.

To encourage discussion on the homework assignment or previous class meeting, ask in a way that acknowledges it is natural for students to have questions. Queries such as "Which problems gave you the most trouble?" or "At which points in the homework did you feel confused about how to proceed?" open the door for students to pursue curiosities – even if they ultimately came to a correct solution for each homework problem. An alternative is to have students report the troubling homework problems through an email prior to class, on a slip of paper turned in as they enter class, or on the board before class starts. This allows you to select the most representative problems – or specific skills – for the class discussion and may assist less-vocal students in asking for help. You could have students include any questions they may have from the previous class meeting as well. Encourage them to identify something specific that caused confusion, such as a certain step in an example, the use of a particular technique in an exercise, or the definition of a term. While you may not have sufficient time to take many homework questions in class, or you may prefer to handle these in office hours, it is important to address major issues before moving forward. Get a sense of where students may be having trouble and address those points in the lesson ahead whenever possible.

In addition to fielding questions, there are opening activities you can employ prior to the start of the next lesson. One option is to ask students to help you review the main topics you covered in the last class, by first working individually to recall highlights, then in pairs, and finally sharing their results with the class (Felder & Brent, 2016: 129). If you have already taken homework questions, this review may yield a natural progression into the topic for the current lesson. Another option is to give a short ungraded quiz or exercise on recent or prerequisite material needed in the day's lesson. The purpose is to foster retrieval rather than test whether students have mastered the material. Once students have had sufficient time to complete the exercise you can discuss the correct answer as a class. This type of activity is referred to as a *minute paper* (or *one-minute paper*) and can be used in a variety of ways to give students a quick test of current skills and comprehension or offer an opportunity for questions.

During class discussion, create frequent opportunities for memory retrieval. In *Small Teaching: Everyday Lessons from the Science of Learning*, James Lang notes that when asking these retrieval-based questions it is important not to allow students to look up the answers, as the research has shown it is less effective to review content than to attempt to recall it (2016: 28). In fact, your students' potential struggle to remember material or their initial failure to correctly answer these questions could increase their ability to comprehend and recall the content in the end. In *Make It Stick: The Science of Successful Learning*, the authors assert that "[u]nsuccessful attempts to solve a problem encourage deep processing of the answer when it is later supplied, creating fertile ground for its encoding, in a way that simply reading the answer cannot" (Brown, Roediger, & McDaniel, 2014: 88). A benefit of a written exercise is that it allows students who may not arrive at answers as quickly as others to have the same precious practice retrieving the material. If you prefer to just take answers orally, consider taking answers by hand-raise after allotting a few minutes or so for the class to think about the question. You can also consider using a mobile polling system, in which students use their phones or laptops to enter responses, or student response systems known as clickers (both discussed in Chapter 7).

The nature of mathematics lends itself well to infusing these moments of retrieval throughout the discussion. As you begin new material, consider asking students to direct you whenever possible. You might ask, "What might I do next?" in a relatively basic algebraic or computational exercise or ask them to predict possible correct avenues to solve a novel problem. Asking your students to make predictions can increase their ability to learn the concepts involved and retain the knowledge (Lang, 2016: 34). As Lang explains, this approach provides a natural way to get your students' attention and prime them for processing the information which follows.

> Asking someone to make a prediction represents a very simple route to raising curiosity and hence represents a very simple route to stimulating the brains of our students and preparing them for their learning.
>
> (2016: 44)

Even incorrect guesses are beneficial, provided they are corrected within a short time frame, and students gain from reflecting on why they made a prediction and why it was correct or incorrect (Lang, 2016: 38–9, 43). By asking students to explain why they propose a certain strategy to solve a problem, you can provide them an opportunity to reflect on their prediction. Ending an example or class discussion by reiterating the correct strategy and rationale will provide any needed clarification.

In order to make predictions, students must have some relevant background knowledge upon which you can draw. If they are struggling to make a guess, you can try to draw this out of them with leading questions, such as "Does this remind you of another type of problem we did recently? What did we do last class that was similar?" and "What is different here?" or "Why won't that approach work now?" Drawing a parallel to a problem they understand makes the more complicated problem more approachable and helps you to distinguish why it is different and may require another tactic. This integrates the various topics you have discussed, intertwining with previous material rather than adding on top of it. As the authors of *Make It Stick* explain, learners who can make such connections improve comprehension and recall.

> The more you can explain about the way your new learning relates to your prior knowledge, the stronger your grasp of the new material will be, and the more connections you create that will help you remember it later.
>
> (Brown, Roediger, & McDaniel, 2014: 5)

It is important to showcase *your* thinking when you are demonstrating the solution to a problem. Being *mathematical* in front of the class improves the likelihood of your students learning these practices and developing mathematical thinking (Badger et al., 2012: 44). For example, suppose you are faced with an integral for which the integration technique must be determined, and you ask your class for suggestions as to what might work. If you just grab one that works and run with it, you miss an opportunity to expose your thought process when you approach such a problem. Instead, you can ask why the proposed method stands out as the method which might best apply. If your class is not yet able to discern which method is appropriate and which will fail, then walking them through your thinking with each suggested approach (or each approach you have taught thus far) gives them a better chance to understand how to analyze an integrand in the future. Research indicates that we can help students develop their ability to select and evaluate thinking strategies by making our thinking visible (National Research Council, 2003: 21).

Write important aspects of problems on the board, including those which address the thought process behind the approaches utilized or discarded. While professors in advanced math classes may typically write out formal mathematics and examples, they frequently provide the informal content orally (Fukawa-Connelly, Weber, & Mejía-Ramos, 2017: 578). Students may not properly assess important points in a lecture, even when the professor pointedly emphasizes them (Lew et al., 2016) and typically record notes written on the board but not oral content (Fukawa-Connelly, Weber, & Mejía-Ramos, 2017: 578). Thus, providing a visual summary or checklist of the approach may be necessary for your students to more fully digest the strategies employed.

Your students will not learn as much material, nor learn the material as deeply, if you purely lecture to them. Students who are only exposed to the material through lecture are less likely to remember the skills in a later course, to be able to apply the material from the course, and to develop problem-solving skills (Fink, 2003: 3). After working through a concept together, take time to let students switch gears in some way which allows them time to try the skills without you posing leading questions. Having the class work through short problem sets or worksheets can be beneficial, especially when they have the opportunity to ask you questions individually and to confer with peers.

Consider breaking up class with designated blocks of time when students try to complete the problems on their own, followed by time to work with a partner or small group while you circulate to take questions. When several students realize they all have the same question, they are more emboldened to ask it. If a small group thinks an answer seems reasonable, they are more likely to put it forth. This provides you with an opportunity to reinforce or correct the group's approach. Discussing differing solutions that arise benefits all students, especially if the class can analyze why one solution is accurate and another is not. Having students perform the desired skills and reflect on the topics provides an active learning environment, which improves and deepens their education (Fink, 2003: 104–6).

If you circulate while students are working on problems, strike a balance between being available to help them and appearing to hover over them looking for errors. You do not want students to avoid writing anything down for fear you may tell them it is wrong. As you walk through the class, you can offer to assist anyone who may not be writing or who looks frustrated. If the classroom is not particularly suited to circulating, then scan the class looking for anyone who seems to want assistance and encourage students to raise their hand if they run into any trouble.

As you select examples, avoid rote practice of the same type of problem. Consider varying examples and "interleaving" material. For example, to provide varied practice of the skill of solving a linear equation, you would give students linear equations in different forms. By doing so, students need to generalize the concept of isolating a variable rather than memorizing an ordered list of steps. *Interleaving* material involves interwoven exposure to multiple concepts or skills, rather than mastering individual skills in a linear way. For example, to interleave two skills, you would teach one topic and, before students have practiced sufficiently to master the skill, move on to a second topic. Before the second skill is mastered, you would resume practice on the first. An example in integral calculus would be to teach an integration technique, practice some examples, teach a second technique and practice a bit, then provide students with a mix of these problems to practice. This approach allows students to learn the invaluable ability to discriminate between different kinds of problems and select the appropriate strategies to solve them (Brown, Roediger, & McDaniel, 2014: 65).

At the end of class, you will likely ask if anyone has any questions, but do not be surprised if most students are more focused on getting to the next class (or lunch) than continuing to discuss mathematics with you. You could use the last few minutes of class to have students break into pairs or small groups to come up with questions, then share with the class (Felder & Brent, 2016: 129). Alternatively, you might opt to have students work individually in the last minute or so to jot down the important ideas in recent lessons, discuss the relationship between topics, or any remaining questions they have (Fink, 2003: 117). If they hand this minute paper in to you when they leave class (sometimes referred to as an *exit paper* in this instance), you will get a snapshot of their immediate takeaway and a sense of any confusions and misconceptions you want to address before continuing. This practice promotes engagement though the required reflection on the material learned, demonstration of skill, or conceptual description (MAA, 2018: 7).

Another opportunity for a memory retrieval exercise is by closing class with a quiz on the day's discussion (Lang, 2016: 31). This quiz need not be graded but, as with the opening quiz, will be most useful if completed without looking up answers or closely following an example. In *Make It Stick: The*

Science of Successful Learning, the authors note that "[a] single, simple quiz after reading a text or hearing a lecture produces better learning than rereading the text or reviewing lecture notes" (Brown, Roediger, & McDaniel, 2014: 3). You could implement a polling system for this exercise, if you prefer.

While you may not have time to give a *graded* quiz every class period, you should consider including one on a weekly basis. Lang notes that "the more students practice retrieval, the better they learn. Frequency matters. The easiest way to implement frequent practice is through regular quizzing. … Give quizzes at least once a week, and don't hesitate to give them every class" (2016: 30). Consider your ability to grade papers in a timely manner, preferably by the next meeting, when determining the frequency of graded quizzes.

You may face limitations on the format and style of your in-class activities, due to available time for administering and/or grading them, course coordination restrictions, or classroom constraints, but offer opportunities for students to engage, reflect, and learn from errors as frequently as possible. Students need such *formative* assessments, from which they are able to receive feedback on their performance and learn how to improve, as often as time allows.

My students have regularly reported on evaluations that they appreciate that my courses have a routine. Each class has an opportunity for questions and practice within the discussion. A typical week has a quiz on the same day of the week and exams are always preceded by a review day that students deem useful. This is not to say that students don't enjoy the occasional change (no quiz this week!), but they appreciate the dependability that a reliable daily and weekly schedule can offer. It is also important for individual classes to have some variety, such as in the types and frequencies of activities, as this increases the likelihood of holding students' attention (Felder & Brent, 2016: 75). Some may be more handout-intensive while others might be more motivated by discussion.

Quick Glance: Typical Day Overview

- At the start of class, investigate questions the class may have from the previous lesson and assignment and motivate the discussion ahead.
- Consider opening with a minute paper (or longer exercise) which assesses current comprehension, prompts retrieval, or demonstrates a need for the material to be covered.

- Ask students to make suggestions and predictions during class to better prime them to receive and retain information.
- Draw parallels to prior material, whenever possible. Ask your students to draw parallels or start with the older material and ask them to conjecture how this might extend to the current material. Note: It is important to assess whether your students recall and understand such previously covered content.
- Showcase your thinking and have students reflect on their own strategies. Highlight the thought process for a problem *in writing*.
- Provide time for students to try problems on their own and/or in groups. Give them varied exercises and interleave material whenever possible.
- Consider closing class with an activity which invites questions or seeks a summary of the recent material.

The Importance of Practice

Mastery comes with practice and engagement, not a thorough exposition from you. Anyone who has helped children learn how to tie their shoes knows that it cannot be done simply by having them watch you do it. If so, no lesson would ever be necessary as you have been trying their shoes right before their eyes, possibly multiple times a day, for *years* before they have the dexterity to try for themselves. They must go through the motions on their own to learn them, and they will most likely fail for a while before it clicks.

It is easy for students to have the misconception that the problems are easy if they only witness you solving them with no difficulty. Giving students a chance to see where they might stumble will help them formulate questions and motivate them to give the homework proper attention. One or two problems at a time is generally sufficient for students to ascertain whether they are having significant difficulty and ask some initial questions. After you have worked through a variety of examples together, an in-class handout with a mixture of problems, rather than the gradual increase in difficulty done initially, may help further solidify their knowledge. The more that you can interact with students as they practice, the greater your ability is to monitor their understanding and clarify misconceptions.

You may not have time for students to work more than a handful of problems each class, depending on the ability level of your students and the length of the class period, but whatever practice time you can give them will be beneficial. David Sousa, an educational neuroscience consultant who has authored over a dozen books on the use of brain research in education, states: "Practice does not make perfect; it makes *permanent*. Practice allows the learner to use the newly learned skill in a new situation with sufficient accuracy so that the learner will correctly remember it" (2011: 133). If you can offer some practice before students leave the room, you improve their ability to remember it correctly when they return to the concept in their assignments.

Perhaps you feel homework should provide sufficient opportunity for practice or that you do not have time to let students work problems in class. Let me offer the following story told in a sociology course my senior year of university. The professor's young daughter routinely came in after school, threw her bag and coat on the floor and left the front door open. Every day her father would call her over and remind her that she needed to her hang up her things and shut the door, but her habit continued. She was apologetic and acknowledged what she was supposed to do, but could not remember the next day. One day, he met her at the bus and as they walked up to the house, he *asked* her what she would do as she went inside. She hung up her bag and shut the door. He did this with her every day for a week or so, after which she independently completed these tasks without a reminder. I heard this simple story in a rather sleepy course over 20 years ago, but I remember it vividly because of two important principles that it illustrates.

If we leave students to attempt and practice emerging skills for the first time in a homework assignment, with little feedback on whether their approach or steps are accurate, we run the risk of creating a bad habit or misconception at the onset.

> Practice that takes place away from the presence of an instructor can become a breeding ground for overlearning, mindless repetition, and the development of wrong or poor habits. Practice that takes place with the benefit of your presence and feedback has potential to create more powerful learning.
>
> (Lang, 2016: 84)

The opportunity to "walk to the door" with our students, sharpening their focus on strategy as they make their way through initial problems, allows them to form good habits.

The other important concept is that reporting an error after-the-fact is less effective than avoiding the error by encouraging careful forethought. The father reported weeks of correcting his daughter upon finding the door open and her things on the floor to no avail. A single week of guiding her to think purposefully as she entered the home made it possible for her to independently make the correct choices. Imagine the frustration avoided for both father and daughter had this guidance occurred the first week of school.

Pace

I once had a student remark after class that he had never seen anyone who could write as fast as I could on the board. He said it good-naturedly and almost with an air of admiration, but I of course understood he was having difficulty keeping up with me. I have always tried to write quickly so that I can move away from a completed statement on the board and allow students an unobstructed view. I have had to work on taking a step back and letting the class absorb, question, and copy the material in their own time. This is harder when I feel I am summarizing what we have just discussed or setting up a topic I am eager to get into, but students are processing information in both scenarios and need time.

There are two paces that you will need to set for yourself. One is the pace at which the course moves and the other is the pace of your individual lessons. Naturally, these are strongly intertwined. Do not let your desired pace for the course rush a discussion that needs special attention, and in return, do not belabor points that do not require it for the sole purpose of sticking to a schedule. Attempt to maintain a steady course pace, with varying lesson paces, as material dictates.

Course Pace

You may find pace to be a challenging part of any new course you teach. If you have your course schedule provided to you by your department, you may have a very structured guideline to follow. This is helpful when teaching a new course, but also restrictive when you are faced with a class that needs more or less review than your average group of students. If the course outline is left entirely up to you, the latitude to stop every time your students' comprehension and performance are below your expectations may hinder your ability to cover the intended material. Conversely, if you have not provided sufficient opportunities for student engagement, you may cover the content too quickly for them to learn it well.

As discussed in Chapter 2, it is a good idea to plan out your entire course schedule before the semester begins, but remain open to adjusting it, especially the first time through a course. If you are not provided with a course outline, ask your department for syllabi from several professors who have recently taught the course. Having multiple syllabi gives you an opportunity to compare the pace and content of different professors. The commonalities should indicate core learning objectives and the differences might suggest optional content. If you find that you have too little material or too little time, follow up on whether you have made appropriate coverage choices. Marking optional content in your schedule may also help in case you fall behind in your planned coverage.

A key point to realize in setting the pace of your course is that you will almost certainly move faster than some students would like, so consider various facets to gauge the appropriate speed. Your assessment tools may provide insight on the pacing for the period of time which each measures. Poor performances could be a result of insufficient practice opportunities and/or rushed coverage. If the class exceeded expectations, could your coverage and/or assessment have gone deeper?

You may need to make adjustments the first few times you teach a course until ultimately finding the right pace for your typical audience. The goal is to cover the most appropriate material while maximizing students' comprehension and retention. Erring too far on either side of the proper pace for your course will result in a decreased mastery of the material for your class. If you move extremely slowly, undoubtedly in an attempt to make sure that *everyone* understands *everything*, your class may only learn a handful of tools. Students will likely be bored and unprepared for the next course. If you move too quickly, then many students may learn the material on a surface level, but most have not had precious practice time to achieve any level of mastery or sophistication. They, too, will be unprepared to move to the next course, and will likely have developed a frustration with the subject or a false sense of mastery.

Lesson Pace

During class, you can get immediate feedback on pacing from watching your students. Does anyone seem to be listening to what you are saying or is everyone madly writing? You may have one or two students who appear to be very slow note-takers, or who don't start writing until you stop talking, but you should have the majority of the class with you appearing ready to continue.

The more you alternate lecture with moments of active learning such as students working on problems, guiding you to solutions within lecture, and offering input to build the concepts in lecture, the slower notes appear on the board. Thus, these practices not only increase learning in a fundamental way, but also ease the student's task of notetaking. Utilizing them will moderate the pace of your lecture as well as give students a chance to discover where they might have questions before they leave the class and begin the assignment.

A key component to asking for your students' input is waiting long enough for them to answer! This seems obvious, but a few seconds can feel like forever when you expect students to immediately respond to a question. Research has shown that instructors wait less than 1.5 seconds, on average, before providing an answer or proposing another question, when they should wait at least 7 seconds (MAA, 2018: 3). You may not think that waiting a measly 7 seconds will make much impact on your pace, but if students begin to respond and become more engaged, you are less likely to cover material too quickly and you may realize a better pace for your lesson.

When planning your discussion, it is essential to consider the amount of time outside of class that your students will have to work on the material. If you are faced with a topic that does not require excessive exposition time, but *does* need more than the usual amount of attention outside of class, avoid the temptation to cover more material in class. To utilize the rest of the available class time, include activities such as group work or in-class worksheets on the material. You may also want to consider when students may have access to tutors at your institution and when they will have the most access to you through office hours.

Enthusiasm and Motivation

It is important for students to feel you care about the course material. We all have topics that we enjoy more than others, and it can be challenging when you are assigned the same course repeatedly or one that is not your favorite material, but try to bring out the best in it. Acclaimed mathematician George Pólya noted:

> … if you appear bored, the whole class will be bored. … You should do a little acting for the sake of your students who may learn, occasionally, more from your attitudes than from the subject matter presented.
>
> (1965: 101)

This effort can have a big impact on your classroom environment and on individual students. Demonstrating positive emotions, especially in regard to our course subject matter, can increase your students' motivation (Lang, 2016: 107).

Even if you are not planning a career in academia, you presumably have an affinity for the subject and are likely pursuing a math-related career. Let your love of mathematics shine through! There is a correlation between the enthusiasm and caring of instructors with students' motivation and learning (Felder & Brent, 2008), so it serves you and your students well to bring your positive mathematical energy to the classroom. You may be able to motivate students by shining a light on how the skills your class is learning are used in the field to make a positive difference (Lang, 2016, Chapter 7).

Creating a positive social environment and maintaining a social component to your course may increase motivation. You can do this by chatting with students in the few moments before class, demonstrating an interest in their lives and learning their outside interests, and by maintaining a positive tone when you interact (Lang, 2016, Chapter 7). For instance, you can note when students have erred, but implemented a good strategy or improved their skills. Comments such as, "Good approach – this would have worked for…, here we need to try… can you see why?" or "Great improvement! You are almost there – let's discuss how to finish this up," go a long way in conveying that you appreciate the knowledge demonstrated and want to help students improve. Further, by providing students opportunities to learn together and from each other during class may boost motivation (Lang, 2016, Chapter 7).

Students' motivation is improved when course topics are personally relevant or interesting and the difficulty level is appropriate (National Research Council, 2003: 21). Learning your students' interests, determining their existing skill set, and weaving in appropriate applications will be useful in this regard. Similarly, if you have allowed students to make choices in regard to projects, they have the ability to apply material in a realm that matters to them.

A recurring example or theme may create interest in your course topic. Keith Nabb and Jaclyn Murawska have described using real data from a Corvette traveling down a racetrack to determine how long it took the car to reach a velocity of 88-feet per second. This scenario can be used to motivate discussion on an array of calculus topics, including instantaneous velocity, limits, and the Mean Value Theorem (Murawska & Nabb, 2015), as well as the Intermediate Value Theorem and the Fundamental Theorem of Calculus (Nabb & Murawska, 2019). Returning to the same example may pique interest in a way that novel examples for each new theorem may not. They quote a student as having said, "Now when I see something new, I kind of wonder, what does this mean for the speeding car?" (Nabb & Murawska, 2019). This type of curiosity and desire to understand and apply the theory is exactly what we would like to foster in our students.

There are a variety of aspects affecting your students' motivation which are already in place on the first day of class. Some individuals will be driven by the desire to earn a good grade or simply to achieve mastery of the material. But others, while potentially sharing that same desire, may not feel the same motivation. Students must not only value learning or performing well, they must feel they can succeed in doing so (Ambrose et al., 2010: 76). Students who have faced repeated failure in math *and* attribute that failure to their intelligence are likely to feel decreased motivation to try to learn new material. Students who attribute their success or failure to the amount of work they have done and the quality of their study habits will be more likely to be more motivated. When a student views success as determined by innate intelligence, they have what is referred to as a *fixed mindset*. Students who see intelligence as something which can be improved are said to have a *growth mindset*. Students with a fixed mindset, even those who believe they are smart, have decreased motivation to face challenges and are more inclined to give up easily, while those with a growth mindset are more persistent (Boaler, 2016: 6–7).

The good news is that mindsets can be changed. If students can come to view their ability as something which they can improve through sound strategies, good effort, and assistance from others, then they can be more resilient when faced with challenges (Yeager & Dweck, 2012: 306). Making a sincere effort to draw a distinction between students' current abilities and their aptitude for success, through appropriate changes in how they approach the course, may have a meaningful impact on motivation. The way that you approach mathematics instruction can send a fixed- or growth-mindset message, as well. Growth-mindset instruction offers frequent feedback, praise for reasoning, and multidimensional assessment, while fixed mindset instruction may include limited feedback and be more focused on accuracy, rules, procedures, and memorization (Sun, 2018).

Encouragement and Suggestions for Improved Performance

When teaching lower-level material, be aware that you can expect a wide variety of ability and confidence levels. In these courses, you will most likely encounter a number of students who have struggled with math for a while. If you are teaching mathematics at the undergraduate level, then repeated failure in math is something you have not likely experienced personally. If you have done any mathematical research, you may understand some level of frustration, but you are fortunate to have had a background that has given you confidence and a reasonable *hope* that you will eventually find the solution you seek. As discussed in the previous section, students who have faced repeated struggles and failure in math may have lost motivation. They may have been stripped of that valuable sense of hope. Offering encouragement and support shows your students the respect and sensitivity that they deserve and may help to restore some motivation.

Invite the class frequently to attend your office hours and visit your campus' tutoring center with any questions they may have. It not only reminds students of these opportunities, but also alerts them to your interest in offering additional support. You might describe what a visit to your office is like and let students know they could come with a classmate or two if they prefer.

If a student approaches you with concerns about performance, there are a number of areas you can assess in order to make useful suggestions. Pursue questions that provide insight as to where there is an opportunity for concrete actions the student can take to improve comprehension or performance. Some areas to consider discussing:

- What the student feels is going well and how the student is struggling
- The student's typical experience during the class discussion
- Whether the student feels comfortable asking questions in class or following up after the class discussion
- The quality of the student's notes – This can be highly personalized, but you may be able to offer suggestions for improvement or assistance if the student is struggling to leave class with adequate notes.
- How the student completes homework – Is the student following notes closely, but not understanding steps even by the end of the assignment? Is the student working with a tutor or friend who is possibly doing too much of the critical thinking? Is the student starting the assignment just before it is due, leaving little time to get help?
- How the student prepares for quizzes or tests – Does it mimic the testing experience?
- Demands on the student outside of your class.

Keep in mind that the student has come to you for help and keep the tone positive and supportive. In trying to determine any deficiency in the student's approach, make it clear that you are working together to determine a potential solution rather than give any sense you are trying to find fault.

When a student has come to office hours to discuss frustration over poor course performance, it may help to relate an experience to them. Explain that you understand such dissatisfaction by explaining how you struggled in some endeavor, whether it was academic or not. Challenges found in other subjects, sports, musical pursuits, and artistic endeavors may provide the common ground you seek.

Approachability

Inherent in being able to effectively assist and encourage your students is that they feel that they can come to you with questions and seek your support when they feel discouraged. Being friendly and supportive of questions in class will encourage your students to seek help both in the class and in office hours. A bright smile for the wary or timid student at your office door may help dispel the notion that he or she is bothering you. Some students miss the point that office-hour time is specifically designated for them, so reiterate this sentiment at various points during the term. In your efforts to be friendly, remain professional, of course. Students should get the message that you care, but will still require them to work hard, to learn, and to perform well on graded papers.

You may find that simply encouraging students to ask questions during class and office hours does not induce the desired interaction. To nudge students in this direction, you can take some formal steps, such as requiring students to send you an e-mail or stop by your office for a brief visit in the first week of classes (Felder & Brent, 2016: 56). Setting up this communication from the start will make it easier for students to approach you when they have a question on material. You can use this initial e-mail or meeting to ask them to tell you something about themselves and any concerns they may have about the course. Making this first contact personal will demonstrate your interest in them beyond their answers on a paper.

It should be said that there may be days that you do not particularly want to be approached! Perhaps you are tired, overworked, or dealing with a particular stress in your home life. Your external demeanor may reveal to your students a lack of enthusiasm for interaction. This can be costly and if you feel you have given this impression, try to reach out to a particular student or the class as a whole to reopen the door.

Not every struggling student will accept your offers for extra help. Some prefer to work with peers, some may be a little shy, others may not want to make the extra effort, and some may simply find it challenging to meet with you at your available times. There may be deeper issues at play, such as negative interactions with math teachers in the past. A student who is having personal issues may not only struggle with performance but lack motivation to approach you for help. If you find yourself frustrated by a student who may not have responded to your requests to stop by office hours or who is not completing basic requirements of the course, try to remind yourself of these possibilities. Consider the advice offered in *Small Teaching*:

> … whenever you are tempted to come down hard on a student for any reason whatsoever, take a couple of minutes to speculate on the possibility that something in the background of that student's life has triggered emotions that are interfering with their motivation or their learning.
> (Lang, 2016: 114)

If a student feels scolded in an interaction with you, it is unlikely they will relish returning to you for help. When working with students who are not meeting basic course requirements, reminding yourself that a number of factors could be at play may help you keep a positive and encouraging tone.

Acknowledging Hard Work

When you know students are working hard in your courses, tell them so! It helps students to know that even if their efforts are not resulting in the grade they desire in your class, at least you are aware and appreciative of the hard work they are putting into the subject. Often, one of the struggling student's fears is that you will misinterpret a poor performance as an indication of apathy or laziness. Students who want high grades purely for working hard will not be comforted by your acknowledgement of their efforts. To this you can only ask them to recognize that the reality of the working world they are about to enter is that sometimes adults do work hard at their jobs and fail to achieve the desired result of a raise, promotion, or accolade.

Rapport

You are a part of a pivotal time in students' lives and your demeanor and attitude towards them matters. In general, the better your students' relationships with their professors and classmates, the better their undergraduate experience is expected to be (Felder & Brent, 2016: 54). A good rapport with your students is essential for a positive learning environment. Being respectful and considerate will improve your class' ability to learn from you and soothe the tensions for students who struggle. You will hopefully get to know some of your students well, but for those whom you do not, try your best to convey that you care about their experience in your class. The more your students feel welcome and valued, the more likely they are to reach out to you with questions on material.

One of the essential steps is to learn students' names. This is definitely a challenge for me to do quickly, especially with larger classes, but I make it a priority to consistently work at recalling students' names. Taking attendance orally each class, at least until you learn everyone's name, may help. If you have a roster with photos, you can print this and take it to class with you or pull it up on a classroom monitor that you can view privately. While trying to learn names, see if you can scan students before each class to find each student on the roster. After class, consider reviewing the list again. When a student e-mails you a question, you can pull the list out to put a face with the name. If you have papers to hand back in the first couple of weeks, consider asking students to walk up to the front to get it, so that you can focus on each name and face as they walk up to you. This also speeds up the process of handing back papers since you are not yet familiar with where everyone sits. A more overt option for learning names is to have students write their names on tent cards which are placed on their desks for the first few weeks. This allows you to not only call on students right away, but puts names and faces together for you throughout the class.

Try to handle student concerns objectively. For instance, avoid acting defensively if a student complains that a skill you are teaching is useless or that something was confusing in class. In the first case, it may help to briefly explain how a topic may be used in higher mathematics. Convey that for students going farther in math, this skill will be necessary. Even students not pursuing math generally respect your need to properly instruct those who are. In the latter case, try to determine where the student became confused in the lecture. Ask questions that help the two of you pinpoint where the connection was lost.

You can gain useful insight to improve your instruction on a topic from listening to students in office hours explaining when they became confused. Making an effort to understand how a student *heard* what you said in class can help you better craft your lesson and the examples you bring to a future class. It is irrelevant how clear you feel you were in class; the student missed your point and you must help illuminate the concept. If you can also learn from the discussion, everyone wins. It can be surprising to hear a student say, "Oh, I didn't realize that" after stating something in office hours that has already been discussed in class a number of times. Regardless of how frequently the point may have been covered in class, this student just *heard* it. Mission accomplished. Instead of feeling frustrated or defensive, consider whether there is a lesson to be learned. Is there a way to better drive this concept home during class or is this a relatively isolated incident?

Try to convey a distinction between mathematical achievement and intelligence. Sometimes students will assume that since this material is quite natural to you, you think that it should come easily to them. Assuring your student that you are confident that he or she is a talented and dedicated pupil goes a long way in establishing rapport with an individual who may have come to loathe the very subject you are teaching. Be cautious during lectures to avoid describing problems as "easy" or "obvious." Especially in lower-level courses, momentary confusion on a student's part after such a description could lead to ill will.

The way you provide feedback will also affect the rapport with your students. L. Dee Fink, author of *Creating Significant Learning Experiences*, suggests that high quality feedback is not only frequent, immediate, and discriminating, but also "loving," which he describes as being empathetic in delivery (Fink, 2010: 14). Facilitate learning by maintaining a positive and encouraging tone in your comments or corrections. Making a conscious effort to acknowledge good strategy and improvements in students' work, even if there are remaining deficiencies, can promote good rapport.

Occasionally, students may not see your attempts in the light intended. You may read comments in student evaluations from the *same* class which range from "always answered all of students' questions," "really wanted us to learn," and "very supportive and encouraging" to "expected us to teach ourselves," and "very unapproachable." Look to see where the majority of the comments lie. How would the average or the median comment read if you had to construct one? While you should take all students' concerns seriously and consider if there are changes you could make to help everyone feel your good intentions, you cannot predict how every student will perceive and respond to you. Some students will come with unknown academic baggage that will taint their perception of you; all you can do is gain the understanding that your intention was misconstrued and consider ways to clarify this in the future.

Fostering Academic Honesty with Environment

Students cheat for a variety of reasons, including both internal and external factors. The environment that you create in your classroom can play a role in whether a student feels pressured to cheat. In *Cheating Lessons: Learning from Academic Dishonesty*, James Lang lists four aspects of a learning environment that may contribute to students feeling this pressure:

1. An emphasis on *performance*;
2. *high stakes* riding on the outcome;
3. an *extrinsic* motivation for success;
4. a *low expectation of success*.

(2013: 35)

I will discuss in Chapter 4 how your assessment strategies can play a role in the first two items, but note how the last two reinforce the importance of your students' motivation to learn and their belief in the possibility of success.

Bernard Whitley, Jr., and Patricia Keith-Spiegel, authors of *Academic Dishonesty: An Educator's Guide*, describe how creating a positive and supportive climate in the classroom may foster academic integrity. It is imperative that students *perceive* that they are being treated fairly, in terms of both how you interact with them and how you have designed and implement your course. Regardless of your intentions, if there is a perception by your students that you are being unfair, their internal justifications for cheating may increase. Specifically, Whitley and Keith-Spiegel cite the importance of your students viewing you as impartial, respectful, having concern for them as individuals, having integrity in your clarity and consistency with course policies, and behaving with propriety (2002: 43–9). Students must also feel that the assessment and feedback from the course is fair, which will be discussed further in Chapter 4.

Making and Discussing Mistakes

It happens. You will copy a line incorrectly and drop a sign here or there. Every now and then you may even royally mess up a problem. While the latter is often preventable with proper preparation, minor errors are pretty difficult to avoid completely.

When students point out a mistake, respond graciously. *Thank* students for speaking up. This encourages others to contribute in the future and furthers the notion that you are all working together to find solutions. Provided your errors are small and relatively infrequent, your students will not deem you as incompetent, just human.

Occasional errors provide a valuable opportunity to showcase how to learn from mistakes. You can reveal to students how you realized something must be wrong in your work and how to track down the consequences. Perhaps even more importantly, you can destigmatize making a math error. If you choose, you can even tell your class that your brain just grew a little!

> The recent neurological research on the brain and mistakes ... tells us that making a mistake is a very good thing. When we make mistakes, our brains spark and grow. ... The power of mistakes is critical information, as children and adults everywhere often feel terrible when they make a mistake in math. They think it means they are not a math person...
> (Boaler, 2016: 12)

If a student points out a mistake beyond that of a simple copy error, you might ask what signaled something was amiss. If you respond without defensiveness or embarrassment to an error which has been discovered, you can demonstrate how the focus should be on learning how to avoid and detect mistakes rather than using them as a reflection of intelligence or ability.

Encourage students to share and discuss their own mistakes in class! After students have worked on an activity, you might ask students to share an approach that was ultimately unsuccessful. This promotes and elicits student thinking as it conveys an expectation that students share ideas rather than solely correct solutions (Rasmussen et al., 2017). The subsequent analysis of an approach which failed, and when it might be more useful, deepens understanding of the problem at hand as well as the misapplied strategy.

When you make an error in class, it is a judgment call as to whether to fully correct the board work immediately. If an error is minor, such as a dropped sign, it is better to attempt to correct the work before moving on. If the error was discovered late in a complex problem, it may be more confusing and time-consuming to fix everything on the spot. Ask yourself whether it is instructive to redo the work or if the perhaps the next example might sufficiently illustrate the desired skill. In the latter case, very clearly mark where the error occurred on the board and draw a line across the subsequent steps which need revision, but do not completely obscure or erase the content. Students may find it frustrating if you immediately erase everything, since they may have needed time to assess what went wrong. You can follow up with a handout or online post demonstrating the proper solution so that students will have the correct work for reference. Encourage the class to complete the problem before the solution goes up.

Acknowledging Your Own Challenges

So far, I have mainly focused on the students' experience and the many ways you can try to foster the best learning environment for your class. We should acknowledge that, just as our students' personal lives may impact their motivation and attention, our personal lives affect the environment we are able to create on a daily basis. The ups and downs we experience outside the classroom affect our performance.

Like many new faculty, I began my family in my pre-tenure days and this added a number of distractions and challenges. I married two weeks after completing graduate school, then moved twice, bought my first home, and had two children within the next 5 years. My first child was born with an allergy to milk proteins and screamed his way (mostly at night) through the first year of his life. My second child slept effortlessly for the first year of her life, but has spent many, *many* sleepless nights since and thus have my husband and I. As I have only vague recollections of sleeping 8 hours in-a-row, I can attest to the challenges of functioning while highly sleep-deprived. Despite my "pro" status in this arena, I harbor no illusions that my work is never affected. For me, it is the aesthetic components of class that I discussed of enthusiasm and approachability that I suspect suffer the most during the most sleep-deprived stretches.

I do not have the ability to control or eliminate this challenge in my career, but being cognizant of it helps me to strive to limit its impact on my classes. I make an effort to remain highly organized in my preparation for class so that content of my lectures and class discussions will be accurate and well thought-out, even if I feel somewhat foggy on the day of class. I aim to have each lesson prepped two days in advance. This not only builds in catch-up or revision time if needed, but serves the dual purpose that I am prepped should the intervening day of classes be cancelled for weather or illness. I have found it helpful to have a routine to my classes to aid in keeping the student experience smooth even if I am slightly off kilter on a given day. Being aware of your own challenges and limitations will be key in finding ways to soften the effect on your classes.

> **Quick Glance: Class and Office Hours Overview**
> - Seek reflection and questions on the previous lesson and assignment from your students. This could be done at the start of class or through online/email submission prior to class. You may also consider requesting written questions at the end of class to assess what gaps in comprehension may exist after the discussion.
> - Ask for student input frequently. Have students actively engaged in material through in-class activities, in addition to asking for their predictions and suggestions as to how to solve problems.
> - Quiz students often, even if you are not able to give graded quizzes. This process improves comprehension and retention.
> - Give your students frequent opportunities to attempt skills in class, prior to the homework assignment.
> - Monitor pacing closely and adjust as needed. Using in-class activities and student input will not only improve learning and retention but also help to slow your pace if you find you are moving too quickly.
> - Demonstrate enthusiasm for your course material and students.
> - Convey that one's performance level is not predetermined and offer suggestions for concrete changes students can make to improve.
> - Keep a positive tone in communications with students. Avoid making judgements about student effort or motivation.
> - Listen carefully to students' comments, from specific points of confusion which occurred in class to general concerns about your teaching style or approachability.
> - Encourage academic honesty by treating your students with respect and maintaining a fair and consistent application of course policies.
> - Use errors you may make in class as a learning experience to demonstrate how to catch and correct mistakes.
> - Assess the external challenges you face and how they may impact your class. Take steps to proactively minimize the effects.

After Class

As soon as possible after each class ends, make a few notes to yourself. While you may initially focus on a slight error in an example or clumsy exposition of a theorem, try to ask yourself a variety of broad questions to reflect on the class:

> **What worked especially well in the discussion? What fell flat?**
> - What did students struggle with the most?
> - Were students engaged?
> - Do you need to add or remove examples? Should the order of the examples be altered?
> - Did you cover all the intended material? Would you eliminate any material?
> - Did you address all the learning objectives? Should your list of learning objectives be modified?
> - How well did you handle pace? Was class rushed or did it seem to drag?
> - Do you need additional or revised in-class problems or activities? Should any be cut?

You may not know how to fix the concerning elements immediately, but noting the issues for further consideration is important. (See the *Post-Class Reflection Worksheet* in Chapter 6 for a quick method for recording and organizing these components). If you teach this course again, you will appreciate the heads up to something that needs tweaking. This practice not only has you actively attempting to improve your teaching, but is also documenting your efforts to do so, which is discussed further in Chapter 6.

Noting the tools that worked well for you will help you remember and move towards those devices again. This may not only make your experience better the next time you teach this course, but in any number of future courses and for upcoming lessons with the current class. Critically thinking about how well the planned in-class exercises are working may promote immediate changes and improved preparation or design. Reach out to colleagues or superiors if you are unsure how to address a concern.

The few minutes spent answering these questions after a class can be time exceedingly well spent. Honing your approach in class as well as documenting your journey in this way is one of the most time-effective steps you can take to improve your craft and career. Making self-evaluation a habit will serve you and your students well and it can start as simply as these few notes after class.

Assessment

4

Frequent and effective feedback is vital in any mathematics course. The skills you are teaching are frequently interlaced or building upon others. A failure to address errors occurring in the foundational skills will result in your students' inability to successfully complete any later process that relies on them. For instance, a student who learns the process of integration by parts but lacks consistent basic anti differentiation skills is unlikely to successfully complete an integral requiring integration by parts.

Students should have opportunities to try exercises and fail without a significant impact on their final course grades. This can be achieved through in-class problems, where there is no effect on grade at all, and assignments where there is some accountability. The opening and closing activities discussed in *A Typical Class* (Chapter 3) offer additional opportunities to provide this feedback. These each provide *formative assessments*, in which current performance or comprehension is evaluated, deficits are identified, and it is made clear what must be done to improve. They should similarly serve as learning tools for you and your teaching strategy.

Students need to be tested on their abilities prior to an examination in a way that imitates the exam. They may miscalculate the degree to which they have learned the material in an assignment, especially if they were following along with examples discussed in class or working with friends or a tutor as they completed it. They can underestimate the mental nudge they receive when another person affirms they can move forward with a step or alerts them to avoid the wrong path. Frequent quizzes are a useful way for students to understand their level of comprehension and retention and for you to see which errors students are making after the initial feedback from homework and in-class exercises.

In *Small Teaching*, Lang encourages a cumulative component to all major assignments and examinations to promote greater long-term retention of the content (2016, Chapter 3). He does not suggest that an exam towards the end of the course needs to weight new and old content evenly, but offers the example of selecting a third or quarter to be derived from earlier material. Asking students to apply or compare the older content to the new presents a natural way to integrate earlier material into the exam. Furthermore, Lang points out that if you will have this cumulative nature to examinations, then quizzes should also take this form to properly prepare students for this approach and to further aid in this memory retrieval. He recommends that you determine a portion of each quiz which will be consistently designated to earlier material. Revisiting older topics in opening or closing questions and as you discuss new problems can help ready students in recalling and applying the content appropriately.

Inherent in these regular opportunities for students to receive feedback and assessment is that graded papers must be returned promptly. Students' errors should be addressed as soon as possible and a discussion regarding misunderstandings is fundamentally important before any attempt to build on the material. Ideally, papers should be returned and discussed at the next class meeting. If that is not feasible, consider looking through the papers to get a sense of where the class may have struggled and discuss these points right away. Ask the students to report what they found most troubling or confusing and attempt to clarify these points before moving on.

Fostering Academic Honesty with Assessment

In *Fostering Academic Honesty with Environment* (Chapter 2), I discussed how the classroom environment can help to foster or hinder academic integrity. Similarly, the assessment strategy can impact the students' internal justification for cheating. Bernard Whitley, Jr., and Patricia Keith-Spiegel cite the student perception of the workload and fairness of tests as two factors which may affect the motivation to cheat (2002: 50–3). They indicate that an excessive workload or tests which do not seem relevant to the course material, are too difficult or are unclearly worded may make students feel cheating is acceptable. Whitley and Keith-Spiegel note that students need feedback which is prompt and clearly indicates which responses are incorrect and why, and professors need to request and respond to student feedback. Finally, they recommend professors offer multiple, varied assessments which accurately reflect the quality of each student's work, and provide a clearly stated grading scheme for the course in the syllabus, which is strictly adhered to whenever possible. All of these actions support the notion that you are making every effort to treat each student fairly and that you want every student to have the opportunity to succeed.

Recall from the previous section on fostering academic honesty, an emphasis on performance and high-stakes assessments may make students feel pressured into cheating (Lang, 2013: 35). This is not to suggest that holding students to high standards for performance is unreasonable. Rather, one should provide students with the proper preparation for high-stakes assessments, with multiple opportunities for learning when the stakes are still low (Lang, 2013: 132). Examples of ways to offer this preparation include minute papers, in-class exercises and group work, and low-stakes quizzes or assignments. There are two effects of this approach. The first is that students can assess their own learning and adjust their study habits prior to the high-stakes assessment. The emphasis is on *learning* before performance. The second is providing transparency as to *what* you value as an instructor and *how* you will assess whether the students have reached a sufficient level of mastery.

Offering your students a variety of assessment styles may contribute to students feeling they have a decent chance of success, even if they struggle with some forms of assessment used.

> Using a variety of means of evaluating students' progress makes it more likely that students will encounter forms of evaluation that are more comfortable for them. This comfort level will help reduce performance anxiety and the accompanying motivation for academic dishonesty.
> (Whitley & Keith-Spiegel, 2002: 65)

Students who do not fare well on timed work may feel they have "no choice" but to cheat if the only assessments are a couple of timed high-stakes tests. Making attempts to assess students through homework, projects, group work, or board presentations may help students feel they have a fair shot, even if examinations are difficult for them. While quizzes may still be timed and closed-book, they are lower stakes and cover less material, so offer an opportunity for students to demonstrate mastery of select skills. In sufficiently small classes, you may be able to offer the option of an oral examination for students who feel more comfortable talking through a problem.

Homework

Homework assignments are a critical component of any math course, and students must have the opportunity to attempt problems on their own without time constraints. These provide an avenue to assess where students are with skills and to challenge them with more time-consuming and thought-provoking questions. The assessment value of an assignment is determined by its design and the degree to which a student's competency is evaluated.

As I discussed in Chapter 2, using routine problem sets as an assessment of skill or effort is problematic, as it is difficult to determine how much of it has been completed by the student you are attempting to challenge and evaluate, without the assistance of technology or other individuals. Additionally, grading assignments of any great depth or length with close attention to detail often cannot be done in a timely fashion, but timeliness is a vital component of feedback. A grade for completion or effort can be completed quickly, but is not assessing much nor providing critical analysis of the quality of work.

If using homework as an assessment, consider including problems which require reflection and application. Problems such as true/false with a justification, describing a relationship, explaining a thought process, and novel applications encourage independent thought from your students, rather than rote execution of rules, following of class examples, or using internet searches. These problems engage students more deeply and will give you greater insight into the level and quality of their comprehension than a typical set of basic exercises. Regardless of whether homework is used as a formal assessment, it is a necessary learning tool for any course.

Quizzes

Quizzes play an important role in providing students a pre-exam opportunity to evaluate where they are with the material and observe the manner in which you will assess their knowledge. They also allow you to determine where the class is as a whole and identify their misconceptions and errors. A class that seems to be following along, participating, and answering your questions well, and generally in-tune with the material can sometimes surprise you with a poor performance on a quiz.

The structure of your quiz should be similar to your examination. Use the same format and phrasing you have used in class and that would appear on a test. If the wording differs slightly from the textbook's directions or terminology, you may want to include a parenthetical reminder of analogous terms. If students find any instructions or questions confusing, clarifications can be made prior to the exam, including which terms will be used on the test. Additionally, using the same format on an exam that you used on the quiz sends the message that the test is just an extension of the quizzes they have already completed.

Failing to alert and prepare your students for a shift in assessment will lead to a *lack* of assessment, because students will study and practice improperly. The material covered on your quizzes should provide the basis for test questions which may require deeper application or a higher level of proficiency. When you are not asking anything (drastically) new or different from the quiz, the exam serves the purpose of allowing students the opportunity to demonstrate the mastery they have achieved on a collection of material, including corrections of prior misconceptions, and potentially demonstrate connections. If your quizzes only tested basic skills and your exam will require significantly more independent and applied thought, you must create a bridge. You could have quizzes gradually transition from testing basic skills to assessing novel applications or offer in-class practice on how to apply a skill in a novel way. If you do not alert students to a significant shift in assessment style *and* offer them some manner to prepare for it, you are not providing them the opportunity to perform their best, nor is your test likely to accurately assess their full abilities.

When you grade a quiz, clearly identify critical errors but do not overwhelm your students with excessive feedback. If you try to correct every subsequent error on students' papers, they may lose sight of where the essential deficiencies occurred.

> … too much feedback tends to overwhelm students and fails to communicate which aspects of their performance deviate most from the goal and where they should focus their future efforts.
>
> (Ambrose et al., 2010: 140)

Circling incorrect work and making a note in the margin of the essential error will be more useful to a student than a mass of red ink and corrections scribbled in. You can provide partial or complete solutions or opt to provide final answers, asking students to follow up in office hours if they are ultimately unable to achieve the correct answer.

Ultimately, the feedback on the essentials is the most important role of the quiz for both you and your students. If you learn that the students are struggling with a basic element of the material, it should affect your upcoming lessons. Perhaps more discussion or practice of the concept is required, or you may need to adapt to your classroom strategy more generally. You might alter intended upcoming examples, working through details more slowly and with calls for them to instruct you through steps, and/or employ more group work. Consider whether you need to implement improved in-class exercises to better identify and address misconceptions early.

You should also evaluate how well your quiz format worked. Were you able to assess the intended skills? Did the structure cause any grading headaches? Were students confused by the directions or did they fail to demonstrate the desired amount of supporting work? Paying attention to these aspects may assist you in writing a better exam.

Preparing Students to Take an Exam

Students appreciate a review guide. Sometimes it may seem obvious to you what the students should study – everything, right?! Even if it is true that everything is fair game for the examination, it is helpful to categorize the material. If students are just skimming through homework assignments, in an attempt to study "everything," they may be missing the distinction between the different problems.

If you listed the core skills and concepts you wanted students to learn prior to writing your lectures, this is a great place to start when drafting a study guide. Look back at the list you created for the various topics and the problems you thought would illustrate knowledge (i.e. your desired learning outcomes). If you want students to review it all, then you can start creating a review sheet from this list. This detailed review sheet can be an excellent starting point for creating the exam. If you prefer to hone the sheet more precisely to the topics you plan to put on the examination, you may want to get a working rough draft of the exam prior to constructing the review guide. Whichever order you prefer, I recommend having a solid rough draft of your exam before distributing the review guide to ensure that the two are in sync.

You can (and should) provide specific indications of the material to be tested without revealing the precise contents of the exam. Here is a sample portion of a review guide for a final examination in an integral calculus course of mine. On a previous page of the review sheet, I stated the general policies for the examination (no phones, graphing calculators, or leaving the room, etc.), listed the specific sections in our text that were covered or eliminated on the final, and noted the relevant problems available in the chapter review sections. On the following sheet, I provided a more specific description of skills expected on the test.

I have provided a specific list of skills that I want students to know and yet, in no way, have I given away the exam contents. Having students help you create this list is also a useful classroom exercise, in which you can inject

---

The final has three major components: Differentiation, Integration, and Applications.

Differentiation
You should be able to:
- Take the derivative of functions involving $a^{u(x)}$, $e^{u(x)}$, $\log_a(u(x))$, or $\ln(u(x))$.
- Take the derivative of functions involving $f(x)^{g(x)}$.
- Find derivatives of functions involving trigonometric functions.

Integration
You should be able to:
- Find the indefinite or definite integral of a function using the appropriate integration technique.

- Identify when and how to perform the following techniques. What are indicators in an integrand that might signal the use of each technique?
  - u-Substitution
  - Polynomial Division
  - Integration by Parts
  - Partial Fractions
  - Trigonometric Integrals
  - Improper Integrals

Applications
You should be able to:
- Set up the integral to find the area of the region. This may include:
  - Sketching a region bounded by a function and an axis
  - Sketching a region bounded by two or more functions
  - Determining where functions intersect
  - Determining the most efficient method to find the area (dx or dy)
  - Determining the limits of integration

- Set up the integral to find the volume of the solid of revolution. This may include:
  - Sketching a region bounded by a function and an axis
  - Sketching a region bounded by two or more functions
  - Determining where functions intersect
  - Sketching the solid formed by revolving the region around the x- or y-axis
  - Determining the most efficient method to find the volume (disk/washer or shell)
  - Determining the limits of integration

Figure 4.1 Sample portion of a review guide for a final examination

examples. I have avoided listing specific examples by each topic intentionally. When you list examples, students will invariably say, "So, if I can do *these* problems, I'm good, right?" Um, *maybe*, but probably not. An effective study guide provides a comprehensive list of the types of problems to expect on the exam, rather than a simple list of questions which may appear (Felder & Brent, 2016: 30).

If you provide students access to previous exams or select problems from previous exams, you should clarify major differences that might occur on the upcoming test. As with giving specific examples on the review sheet, you may have students attempting to refine their entire studying down to the problems on the old test. If you are asking students to apply skills in a novel way, providing previous problems which showcase how this might appear is useful. To maintain the novelty, the example provided would need to differ, possibly applying a totally different skill in a similarly novel manner.

Creating a study guide for your students may do more than simply collect and organize the material to be tested. When students understand the expectations and how to achieve a desired grade, they may have decreased test anxiety (Steele & Arth, 1998). Exposing your rubric or how much weight each section or chapter will carry may help to clarify how you will award partial credit or which material may be highlighted.

Review Sessions

Review sessions can take many forms, but an important aspect is that they be interactive. If you simply re-state everything you have been doing, you probably are not helping many students. Alternatively, if you are only taking questions, you may be failing to address core issues or help the majority of the class prepare. Sessions which involve students actively working on problems that you would expect them to complete on an exam allow students to realize there are both areas of comfort and those that need work. These sessions should stimulate questions on the topics on which you want students to focus.

My review sessions include examples similar to planned test questions, problems which evaluate content somewhat differently than on the test, and limited material not on the test, but which I want students to practice. While I omit some test material from the review session, the corresponding skills are clearly listed on the review sheet. When the question "if I can do *these* problems, I'm good, right?" rears its head again, I reiterate that while the review has purposeful overlap with the exam, we have not practiced every skill on the test.

My overall goal with the session is to not to give my students a practice test, but to showcase how questions might appear on the exam and open discussion about what skills they should demonstrate within their work. I give students time to work on one or two problems at a time, and then we go over the exercises as a class. Students have an opportunity to ask about an approach they may have tried or an acceptable form for a final answer. They may also be reminded of a related problem and ask about its chances of appearing on the exam.

Providing your students with clarity on *how* you will test them is as important as indicating on *what* you will test them. In *The Spark of Learning*, Sarah Cavanagh suggests exposing students to your testing style prior to the exam is one avenue by which you can reduce test anxiety (2016: 188). Ideally, you have incorporated this style into at least some of your in-class exercises and quizzes, but the review session will offer an additional opportunity to alert students if the test format will differ from these. For each question that you propose to students as a method of review, consider whether it represents how you would ask students to demonstrate the skill or knowledge required on your exam.

You might add a little fun to these review problems by turning the session into a game. You can have students work individually or in small teams and have students present correct solutions to the class. You could offer a nominal prize or a small extra credit incentive, but the game itself can infuse a little energy into the test preparation. It is an opportunity to create positivity and a touch of light-heartedness going into the exam.

Pros and Cons of a Selection of Test Prep Options

Test Prep Activity	Pro	Con/Concern	Things to Consider
Review sheet	Clarifies the material to be covered on the test	Unintentional omissions or vague statements could lead to insufficient preparation and ill will. If specific examples are listed, students may only study those problems.	Be sufficiently detailed as to be useful in directing students' preparation, but carefully weigh the use of specific examples.

(continued)

(continued)

Test Prep Activity	Pro	Con/Concern	Things to Consider
Review sessions	Students are able to ask questions about problems they have been unable to complete successfully and seek clarification on general exam topics.	Students may restrict studying to the topics which arise in this session.	Clarify whether important topics have not been discussed.
Review problems (new)	Allows additional practice	Students may restrict their study to completing only these problems.	Include sufficiently difficult problems as to provide examples indicating the level of mastery and application expected on the examination.
Review problems from previous exam questions	Allows additional practice and an opportunity for students to see the style of the exam and the level at which they will be tested	Students may restrict their study to completing only these problems. If the problems are similar to the upcoming exam and intended to test students' ability to apply material in a novel setting, some quality in assessment of this ability is lost.	If testing application of skills in novel settings, consider using examples that apply different skills than those which will be tested in this way on the current exam.
Providing previous exams	Allows additional practice and an opportunity for students to see the style of the exam and the level at which they will be tested	Students may assume the upcoming test will be nearly identical to the one(s) provided for additional practice. May lose some quality of assessment for testing application of skills in novel settings	See above. Point out material or types of problems not covered on the old exam which might appear on the upcoming exam.

Writing an Exam

When you write an exam, consider which of your learning objectives you value the most. I prefer to start the process with a rough draft of my review sheet, in which I list the topics and skills we have learned. In choosing test questions, I start with the most critical skills/concepts in each section and build from there. In selecting your test questions, aim to include a variety of difficulty levels. Consider asking a few problems that allow students to demonstrate fundamental skills, preferably early in the exam. This is beneficial because the weaker students have an opportunity to exhibit knowledge and ability and all students can gain confidence and momentum when the test begins with a comfortable problem. To assess the depth and breadth of your students' knowledge, you will naturally need to include moderate and advanced questions. You can have these different skill levels tested in distinct problems or as parts within the same question, but be careful of setting up problems in which each subsequent part depends on the answer from the one before. This could lead to a grading headache, as well as potentially affecting your students' ability to demonstrate their competency in later portions of the problem if they stumbled earlier.

Consider reviewing L. Dee Fink's Taxonomy from Chapter 1 as you formulate how to assess the skills and concepts you have covered. Can you ask students to make connections between materials, tackle new applications, or engage in reflection? Consider ways that you can approach material beyond basic calculation. If these deeper questions are appearing on the test, the students should have had some preparation and exposure to approaching these *types* of problems prior to the test. While you should feel free to ask students to apply or discuss material in a novel way, you are unlikely to assess their true abilities in this realm if you have never fostered this type of thinking in class, assignments, or quizzes. Students may feel they were not given a fair chance to succeed if a new question format weighs heavily on a test.

When you have collected the questions you would most like to include, pause to evaluate the length. You should be able to finish your exam in less than one-fifth to one-third the time that your students will have to complete it (Felder & Brent, 2016: 169). Consider aiming for the goal of one-fifth. In lower-level courses, students may work more slowly than you initially expect. In upper-level courses, your problems will be more complex, requiring more analysis and thought throughout the exam. Providing ample time for students to complete their work is one way to reduce test anxiety (Cavanagh, 2016: 185–6) and increases the opportunity for students to review their work for errors.

Once you have a proposed rough draft, take the test yourself. By carefully working through your test in advance, you may catch errors or unintended complications within a problem. You can assess whether the layout on the page works well for providing supporting work and whether you have allotted sufficient space. You might also note that you have tested the same skill several times and find you can substitute a different problem or two to better balance the material. If you have already written a review sheet for the test, revisit it to ensure you have accurately described the material which is included. You may opt to alert students to omitted topics, especially if you want to direct their focus to other material. Determine if there are any additions or clarifications you should make prior to distributing the review sheet.

When you assign the points allotted to each problem, take into account that you will want to be able to acknowledge a variation in the level of success students have within their solutions. Create a detailed key which specifies how points are allotted before administering the test (Felder & Brent, 2016: 172) and think about how you will grade students' answers. You may decide to alter your instructions or break the problem into parts, to clarify what you will expect the students to demonstrate. Including your point-values for each problem (or section) on the test may assist students in prioritizing time during the examination. Stating the points for a section instead of individual problems permits you some flexibility post-test if you need to make slight adjustments in your intended allocation. Additionally, specifying the points on individual problems may give away the relative difficulty, revealing which require more advanced skills. One example is a section of integrals requiring a variety of techniques, from simple power rule to integration by parts.

Consider analyzing how your point allotment is distributed over the material covered. Does this make sense in regard to the time spent on the material and/or its significance in the course? Consider communicating the rough breakdown to the students. Suppose that approximately 20% of the test's points are allotted to the foundational sections 1.1 and 1.2, sections 1.3 and 1.4 each comprise 20%, section 1.5 is 25% and section 1.6 is only 15%. Relating this breakdown may help students better understand what you most value and adjust their preparation accordingly. Be sure to stress that these are rough approximations. If you determine while you are grading papers that the rubric needs some adjustments, you will want the room to make modest changes.

Assuming you have an average handwriting size, try to leave one-and-a-half or double the space you would need for the solutions to your problems or have students use bluebooks. Balance your desire to save paper with your students' need to have ample space to experiment slightly or take an indirect path in solving the problem. Having scratch paper on hand during an exam or included in the exam packet is also helpful. Just as advertising the points for a single problem may give away the degree of difficulty, the amount of space provided may hint at this, as well. In a section of problems requiring the same directions (such as differentiation or integration), consider trying to provide similar space for each part, when not too impractical.

Consider asking a colleague to share a test administered in this class and to review your planned test as well. In reading another professor's test, you may find a better way to assess a skill or note topics omitted. Your colleague might discover points of confusion in your instructions or offer suggestions on structure, content, and length.

Final Exams

Throughout the term, the various forms of assessment employed have worked to provide students with continuous, informative feedback and to direct improvements in teaching strategy. A cumulative final represents a *summative* assessment, which depicts the ultimate level of mastery which has been achieved. The final exam is an opportunity to have students connect material and reflect upon their learning. It is also a chance for you to collect data on where students excelled or struggled and consider changes for the future. What further explorations could be done where the students were strong? How might better comprehension be achieved where they were weak?

Quick Glance: Exam Writing Overview

- Include a variety of difficulty levels in the problems you select.
- Use caution in lengthy or multi-part problems which rely on answers from previous parts. If you can separate parts without compromising the quality of assessment, this may make grading significantly easier. This may also allow for better demonstration of skills, preventing early mistakes in students' work from altering or oversimplifying the work required later.
- Consider L. Dee Fink's Taxonomy as a way to provide depth and variety to your test questions.
- Take the test. Assess whether you have achieved a good balance of the skills to be evaluated. Check for errors and unintended complications. Evaluate whether sufficient space has been allotted.

- Verify that you can complete the test in roughly one-fifth the time your class will have.
- Create a detailed key. When creating rubrics for individual problems, strive to distinguish between students who have demonstrated basic skills, those which have demonstrated a higher sophistication with skills or concepts, and those who have exhibited mastery.
- Ask a colleague to share an exam and to review yours.

The Exam Was Too Long!

An easy mistake to make is writing a test that is too long for students to have sufficient time to complete it. If a number of students seem to be *intently* working when the test period ends, you may want to consider whether your test was too long before you start to grade the papers. It should be noted that students may remain after they have had sufficient time to complete the problems for a variety of reasons, from reviewing their work or simply trying to remember how to complete a problem. Flip through the papers to see if many tests seem to be incomplete. If it appears that your class was unable to fully address the problems presented, there are a number of options available to you.

The simplest fix of a test that was too long might be to offer students additional time to work during the next class period, but this presents security issues. It is probably better to announce this opportunity in advance so that all students are aware and attend class. If you do not, any students who are not present will need to be given an opportunity to complete the test and they will have the advantage of preparing in advance.

An alternative to extra time is to eliminate some material or adjust the total points. It is unlikely that all students worked in a linear way, meaning you probably will not be able to just take the last problem or two off the test, but you could consider dropping a set number of points from the test's total or adding a "flat curve" (i.e. an equal number of points is added to everyone's paper). Unfortunately, it is difficult to undo the stressed thinking that can occur when a student is rushed. The impact for some students will be greater than just an incomplete problem. You can consider letting the students choose their best work by handing back the papers with colored pens and telling them to circle their choice for a select number of problems. For instance, you could opt to tell them "choose one of #5 or #7 and one of #8 and #9 to be graded; all other problems will be graded." This gives you more control over what you are assessing but gives students control to select the problems that represent their best work. Other options would include allowing students to do corrections on the entire test, on a preset number of problems, or on a predetermined set of problems.

None of these solutions is ideal, which is why it pays to adhere to the guideline mentioned above of a test which you can complete in one-fifth to one-third the time your students will have. Regardless of your efforts to match the test length to the time available, if the test was too long to provide a proper assessment of your students' abilities, you need to make adjustments in some way. It is unfair to penalize the class because you misjudged the time necessary. Students will appreciate your attempts to make it right, and you do not have to sacrifice standards. Take time to think through what you feel is both fair to students and best maintains the assessment you are seeking.

Pros and Cons of a Selection of Methods for Too-Long Tests

Method	Pro	Con/Concern
Provide additional time during the next class period	Students have an opportunity to finish their work and potentially correct errors they may have made in haste as they rushed to complete the test. No additional work is required on your part to create new test questions. Avoids curving or cutting problems meant to assess skills	This presents security issues, especially if announced in advance. If the additional time is not announced in advance, a student who missed class and completes the test later has an advantage over students who attended class.
Eliminate material	Avoids any security concerns, as you will only use the work completed during the initial sitting	Loss of some assessment is likely. May be difficult to design an elimination that assists students equally. May fail to properly differentiate between levels of mastery or the sophistication of responses. Students who felt rushed may ultimately have done substandard work on problems not included in the eliminated material.

(continued)

(*continued*)

Method	Pro	Con/Concern
Adjust the total number of points or add a flat curve to each student's score	Avoids any security concerns, as you will only use the work completed during the initial sitting All work is graded and students' rank should be preserved.	Students who felt rushed may ultimately have done substandard work on the exam. These students' abilities may not have been fully assessed.
Allow corrections	Students have the opportunity in an untimed setting to fully answer all questions and present their best work.	This presents security issues and may not accurately reflect a student's ability to solve problems independently.

Grading an Exam

As you grade the exams, it is helpful to keep a record of your rubric on your key, so that you can maintain consistency from paper to paper and quickly and accurately answer students' questions about lost points. You may need to add notes to your rubric on unexpected errors that students made and how each affected the points awarded. If you find that the rubric you designed is not working well for a problem, then you probably need to change it and regrade the problem. The guideline you establish should distinguish between students who wrote nothing relevant to the proper solution, provided work evidencing an understanding of some portion of the problem, demonstrated a clear understanding of the problem and made substantial progress towards the solution, and those who solved the problem in its entirety to your satisfaction. While this may not be easy to achieve in low-valued quizzes, you should have sufficient depth on an exam to do this for most problems. Taking the time to grade the problems this way should ultimately help you arrive at a reasonable *distribution* of test grades, even if the overall scores are higher or lower than desired.

Keep in mind the goal of purposeful, not excessive, feedback while you are grading. Mark the critical errors and write brief comments to clarify what approach the student should have used or why a strategy failed. Let students absorb those remarks rather than potentially become overwhelmed deciphering your detailed notations throughout their work. They will gain more from making the correction on their own and retrying the problem. This is not to say that you need to ignore subsequent errors, but focus remarks on the vital points so that the students are clear what they most need to immediately address.

To maintain consistency between papers, it helps to grade a single test page or problem at a time. If you are grading a problem on the second page of a test and have a question as to whether you maintained consistency with similar papers, it is much easier to locate these while all the tests are open to the same page. Similarly, if you decide to alter your rubric, you can more quickly check back through papers you have graded for any necessary adjustments. This approach may affect the format you select. For instance, using blue books can create hassles of locating students work and flipping through lots of pages. Give some thought to how you plan to grade and consider how test format could assist or impede that plan.

> **Tips for Grading Exams**
>
> - Before you start, confirm that it appears students had sufficient time to complete the exam.
> - Write your rubric on your key, if possible, for quick reference.
> - Consider grading one page or one problem at a time.
> - Keep feedback focused and informative, but not excessive.
> - Refer to the rubric frequently to confirm you are maintaining consistency and add notes regarding how you deducted for unexpected errors.
> - If you need to adjust the rubric, review papers that have been graded to make necessary adjustments.

Post-Mortem for You

After the exams are graded, mull over the results before you return the papers to the class. How do you feel about the grade distribution on the exam? Were the average and median scores what you expected? If the grades were either too high or too low, put some thought into why and consider how you can address the concern now or on a future exam.

If the grades are too low or students did not perform as well as desired:

- Was there a particular question which most students missed? If a question was missed nearly universally, determine if the fault may be yours. (Hint: most likely!) Does it seem students misunderstood the question?

Perhaps it was poorly worded or used unfamiliar terminology. Perhaps there was some confusion over how the material would be tested. If so, you should consider removing the question or allow a subsequent quiz on that material to replace the points for the test question. Occasionally, students will be given fair warning that a problem will appear on the test, but they come unprepared to answer it. You might not feel obliged to offer compensation in this case, but at the very least, you should consider how you can better prepare students in the future. Perhaps more opportunities for practice or retrieval were needed in class or on a quiz. Do not hesitate to (confidentially) survey students as to why they missed the question and consider following up with this material on a subsequent quiz or test.

- Did you grade the test too harshly? If your *distribution* of scores is as desired, this suggests you have achieved an assessment in the differences in your students' abilities. In this case, adding a flat curve to assign suitable letter grades may provide a satisfactory solution. If you are not satisfied with the distribution, determine if regrading specific problems might rectify the situation.
- Was the test too difficult? If so, determine how you could alter the next test – or the *preparation* for a similar test – and find a way to make things right for the students. You could add points back or allow them time to re-try and complete the more intricate problems.
- Does it appear the class may not have studied the correct material? How could you have better prepared the students on your review sheet or in your review sessions? If problems were not attempted, you might ask students whether they did not study the concept or blanked. In the latter case, it may not have been your test review, but an absence of retrieval practice leading up to the exam. Could you inject more opportunities for retrieval? Sometimes students will indicate that they ran out of time when the real issue was recalling how to complete the problem. Observing your class while they take the exam, and noting whether students are actively working towards the end, may assist in the distinction.
- Are the students' skills sub-par? Consider whether you need to increase or improve upon your in-class exercises to aid practice and retention. Perhaps your students need better feedback on homework and quizzes.

Determine the most critical points that need to be addressed and put plans to move forward on hold until you can repair any cracks in the foundation.

Consider retesting students on the critical skills and use the retest score in conjunction with the original test. Two options would be to count the higher of the two tests or add a percentage of the additional points earned on the retest to the original test (Felder & Brent, 2016: 173).

If the grades were too high:

- Did you grade too easily? You may be able to identify a problem or two for which a more detailed grading would better help you differentiate between students who demonstrated more complete work or sophisticated thinking. If so, you may be able to resolve the issue by regrading those problems with a more discerning eye.
- Did you make the test too easy? There is not much you can do for now, but consider how you could make the next test more challenging. If you do not provide some increase in challenge, your students may begin to check out and shift their attention to other courses. While you should certainly commend your students on their performance, alert them if you will be increasing the difficulty the next test. You can start challenging the class in quizzes to give them a good expectation of how the next test might be different.

If the grades seem just right, awesome. Keep in mind that each new group of students you encounter may need a slightly different approach or level of difficulty. Even when things go well, you may develop alternatives for the future.

Post-Mortem for Students

After you have carefully assessed your own role in how students performed on the examination, think about how your students should move forward. If students need additional practice on skills that were tested, try to offer specific suggestions for improvement. Help students find additional practice either by generating problem sets or finding some available online. You can make up a sheet with sections for different skills; "those who missed #5 should practice problems, such as ...." Offer to work with students on these extra problems in office hours or a review session.

If students did a great job on the exam, make sure you tell them so. They will appreciate your acknowledgment of rising to the challenges you put forth. Showing appreciation for hard work regardless of outcome is important.

Addressing Instances of Academic Dishonesty

First and foremost, familiarize yourself with your institution's policies and procedures and *disregard any advice offered here which conflicts with those*. Because cases concerning academic dishonesty can involve legal matters, always follow the guidelines put in place by your institution. Failing to do so could violate your students' rights.

You will want to find out how to report incidents of cheating and how to properly address concerns with a student, preferably before you ever encounter a situation in your classroom. If you suspect cheating as you are grading a paper, talk to a couple of colleagues and possibly the chair of the department without revealing the student's identity, before addressing the issue with the student. Sometimes a reasonable explanation for suspicious-looking work may not have occurred to you. When an incident occurs in class (such as discovering a student with a crib sheet), address the matter as quickly and quietly as you can so as to not disturb the other students and ask the potential offender to speak with you after class about setting up a meeting. This protects the student's privacy and gives you an opportunity to seek advice.

If you deem that the occurrence should be further addressed with the student, keep an open mind when you ask for an explanation for whatever concerns you about the submitted work or incident observed in class. It is easy to feel angry or offended when you think a student has cheated in your class, but try to keep in mind the variety of factors that may drive a student to cheat, which may have nothing to do with you or the way you are conducting your course. For instance, recall the aforementioned "extrinsic motivation for success" in Lang's list. A student with demanding parents may feel pressured to get high grades at any cost and respect for you and your course may be of little consideration. As Lang puts it, "[a] student cheats on an assignment or an exam–he does not cheat *on you*" (2013: 220). Strive to have a calm, reasonable conversation in which you listen carefully to the student's explanation. If you need additional time to think over the response, take it. If you believe there is evidence of cheating and the explanation was inadequate, explain to the student your rationale and follow through with the appropriate procedures. When you have an incident which occurs during class, you may have an immediate emotional response, but you must remember to act respectfully. Say as little as possible in front of other students. If you have incorrectly assessed the moment, you will be grateful for reserving your comments for later.

While institutional procedures can be a hassle, avoid handling these instances on your own or ignoring them altogether. Doing so prevents the administration of tracking instances of cheating across the student's courses and may do little to discourage the student from cheating again (Lang, 2013: 223). Additionally, imposing penalties without following the official procedures at your institution could be a violation on your part and may infringe on the student's rights (Whitley & Keith-Spiegel, 2002: 118, 120).

Course Grades

Your course grades must be calculated as specified in the syllabus. Depending on the latitude you have been given, you may have the ability to adjust final numerical course grades *up* to better assign appropriate letter grades. Take care in doing so, so that you are improving the accuracy of the grade reported, not arbitrarily raising grades.

There is a subjective nature to any grading, even mathematics. Each mathematician will personally value different aspects of material and individual problems. Some mathematicians prize conceptual understanding over competency with mechanical skills. Both are important and research has shown that concept-based instruction need not lead to a deficit in skills (Chappell, 2006; Chappell & Killpatrick, 2003). The assessment strategy will differ for colleagues with varying viewpoints how to value each and their course grades reflect different achievements. Whatever your desired learning outcomes may be, it is essential that your grades provide an accurate assessment of how well your students achieved them.

While you may want to reward a diligent student in your course with latitude in his or her grade, you may also be doing your hard-working student a disservice if the student is taking another math course beyond yours. This is especially true in the case of a borderline failing grade. If students need the material in your course to succeed in the next course, you may be setting them up for failure in the future by awarding unearned passing grades.

Being Honest and Objective

It can be difficult to be objective when you work closely all semester with a dedicated student who ultimately does not meet performance standards. You hopefully built a rapport and camaraderie by the end of the semester and you naturally want your student to succeed, but be truthful if a student falls too short of the cut-off point. It is your responsibility to honestly describe your students' abilities (Elbow, 1986: 143). If you feel that you are failing to do this reliably, consider why this might be occurring and what changes you should make to your grade allocation structure or your assessment tools in future courses.

If you have worked extensively with a struggling student, be clear what your standard will be for a passing grade. The student could misinterpret your encouragement and acknowledgement of hard work as an assurance that you will not award an F. It can be challenging to be supportive while reminding a student what level they must achieve, but a misunderstanding in this setting is especially unfortunate and should be avoided.

Extra Credit

Since your course grade should provide an accurate description of a student's proficiency with the course material, any avenue in which you permit a student to raise a grade should reflect an increase in mastery and comprehension. It can be difficult to make extra credit in an undergraduate math class meaningful. Anything beyond the scope of the course will distract the student from the work that needs to be done to master the requisite skills. This could be detrimental to future performance. In all likelihood, the students who will be most successful in such endeavors are not in need of extra credit.

You could allow students to submit additional routine math problems as is sometimes done in high school, but this essentially amounts to awarding extra credit for studying and practicing problems and may not be improving the students' abilities. This might be more appropriately viewed as part of an effort or participation grade. Maryellen Weimer has written multiple books about teaching in the college classroom and has served as an editor of the monthly teaching newsletter *The Teaching Professor*. She draws the distinction between this type of exercise and a more thoughtful extra credit opportunity in a follow-up to her piece *Revisiting Extra Credit Policies*. She asserts that a genuine "second chance" should be designed by the instructor and it should engage a student in "robust" work which has the ability to increase comprehension or develop skill (Weimer, 2011, August).

A common option is to allow students to earn points back on exams by doing corrections. You may want to reserve this practice for the rare occasion when the bulk of a class does poorly on an exam. In this instance, most students may need to spend time revisiting the material so that they can proceed with the course successfully. In addition, these allowances help build morale back up and let the students see that your goal is their learning. Perhaps there was some miscommunication about what to expect on the exam or more in-class practice was needed. Something went amiss, so it is important to discern the issue and attempt to rectify it in a meaningful way.

It is certainly possible that students are learning from mistakes by making corrections, but you may worry that students will lean too heavily on notes, fellow classmates, online calculators, or apps. An option to address this concern is to hold a follow-up short test addendum on the most commonly missed exercises after students have had the opportunity to correct their exams and follow up with questions. This creates a greater motivation for students to understand the corrections they make on a deeper level and reinforces their learning. You could use the addendum score as extra credit towards their examinations instead of grading corrections. This may better represent a proven improvement in the students' abilities than corrections alone. A downside is that a student who already excelled at these problems may be getting little educational value from the exercise.

Weimer describes an alternative option in which a professor allows students to leave the exam with a list of questions which they were unable to answer or in which they lacked confidence (2011, July). Prior to the next class, students could find solutions to those problems through any means other than the professor. These alterations and additions were turned in at the start of the next class and could count towards half the credit missed on the corresponding problems. She notes that students felt this was more beneficial than simply being told the correct answers and it decreased their test anxiety, while the professor felt students had to apply additional thought in determining whether their answers had been correct. With the availability of apps and internet sources, the argument could be made that a student could bring in solutions to a variety of math exam questions having applied no additional thought, but you could modify this approach if you have sufficient time.

Instead of at-home corrections, you could offer an optional, subsequent test consisting of equivalent, non-identical problems. This has the downside of requiring more testing and grading time, and may not reduce test anxiety as much, but maintains better accountability and might more accurately assess improved comprehension. It provides a broader opportunity than the test addendum in allowing students to attempt any type of problem rather than those most frequently missed, but will take longer to administer and grade. An advantage is that in this modification students do not need to take a list of questions out of the exam, which may be undesirable should another section be taking your exam later in the day or if any students were absent.

Instead, you could opt to let students see graded exams, and note the problems they missed, before offering the subsequent test.

If you do not plan to offer any extra credit in your course, alert your students (especially first-year students) to this policy, since many of them are accustomed to having extra credit as a standard option in high school. If you will offer extra credit, clearly outline how students may earn this extra credit and the maximum effect it could have on a student's final grade. Opportunities for extra credit should be made to the entire class or not at all.

Consistency

Your grading should be consistent across all students. When grading papers, keeping a record on your answer key of how you have deducted points assists in maintaining uniformity. This record is also useful when a student inquires about lost points on a problem, especially if some time has passed since you graded the paper. When computing final course grades, you should follow the syllabus and the grades should accurately reflect each student's ability to perform skills without assistance.

5 Challenges and Opportunities within Commonly Taught Courses

Each course you teach will come with its set of challenges and opportunities. In this chapter, I will discuss a selection of standard courses. For the entry-level material, I will address some common issues that might arise, through the use of specific examples. While your course may not cover these exact skills, the examples should illustrate a general teaching concern which extends to the material (or typical audience) for the course. There are also a variety of handouts provided for in-class activities. For the upper-level content, it is assumed that you have gained some experience prior to teaching the related course, so the focus shifts in those sections to general techniques you might employ.

It should be noted that this is not intended to be a comprehensive list and, depending on your institution, your audience may have a differing skill set than is presumed. Due to the variety in learning objectives and desired learning outcomes that might be employed across institutions, I have not suggested what specific content you should deem essential in each course. For guidance on this, you should turn to your department chair and colleagues, as well as the general discussion offered in Chapter 1.

Discussion Sections

If you are a graduate teaching assistant, you may have been assigned to teach a discussion section (or "fourth hour") of a course in which you generally answer homework questions rather than teach new course content. This is a nice way to ease into teaching if you have not taught before, since you will practice presenting math problems and gain insight as to where students get confused, but you may not be responsible for constructing any lessons. These sections present their own set of challenges.

In graduate school, I was once assigned to conduct a discussion section for a differential equations course with an emphasis on physics, which contained material which I had never seen before. The students in the course were quite sharp and the combination of their ability and my lack of exposure was more than a little daunting. For me to succeed in that setting, it was essential that I prepare well. I read each section in the text that the students were covering and completed all of the assigned homework. While this was sufficient for my purposes, if you find you are unsure of the material you are to cover, contact the professor teaching the course to address any questions you may have. You may find it helpful to attend the lecture as well.

In another discussion section I conducted, students were unhappy with the instruction from the professor for the course. They felt quite lost in the class at times and would occasionally look to me to illuminate a week's worth of material. I learned quickly to come to class with a list of relevant formulas, definitions, and other such very brief notes on the major concepts involved in the assignment. Having a high degree of comfort with the material covered in the assignment is extremely important in any discussion section, but especially so if you find the class often requires a significant amount of help on the basic concepts. Be prepared in any case to address common areas of confusion in the material.

Even if the class adores the professor for the course and has learned the material well, you may still face some difficulties in reviewing the assignment. Any variation on your part from the notation or method used by the professor may be met with dismay. Students in the first year or two of undergraduate study often fail to recognize when methods are equivalent and minor differences may confuse them. Check the text to see if the students' book uses the notation and method you would naturally use. If it differs, you may want to check with the professor teaching the course to see what was presented in class. Consider asking students as you start a problem if this notation agrees with what they have seen in class. If students do question your notation or methodology, attempt to switch to what they have seen in class. Your goal is to be cohesive with the professor and to help students learn the material as originally presented whenever possible.

You can learn a lot from teaching discussion sections. Perhaps the most important lesson is to be prepared for a wide variety of questions. Regardless of the environment in the class, the students' skill level, or your comfort with the material, do not underestimate the value of your readiness to handle the specific homework problems, as well as the material in general.

Algebra – Is It Too Late?!

By "algebra," I mean the skills *hopefully* learned in high school algebra courses. Unfortunately, the skills possessed in this area by some high school graduates can be appalling to those teaching mathematics at the undergraduate level. The question, which begs to be asked, is *"Is it too late to learn algebra skills in college?"*

It is the rare undergraduate student who is actually *incapable* of learning this material with the right amount of dedication, time, and materials, paired with the right teacher for that student's learning style. However, students possessing *extremely* sub-standard algebra skills upon reaching college may feel a future in math is unlikely. Therefore, a responsible prioritizing of their workloads may require that they limit the amount of time spent attempting to master skills they struggled with at length in the past. This is especially true when the material in question is not material in the current course but presumed background knowledge.

The reality is that it takes much more work to deprogram incorrect notions about algebra than to learn it correctly in the first place. It can require extensive drill work to accomplish this deprogramming – something you generally do not have time for in the semester. So, we reach the question: *"What do I do as an instructor faced with teaching students who lack basic algebra skills?"*

It depends on the focus of the course at hand. If you are actually teaching a remedial math course, which is *intended* to patch up these holes in students' mathematical backgrounds, then you must go back to basics and go slowly. Remember, it takes time to uproot all of the weeds of misconception and replant the knowledge correctly.

> … it is important for instructors to recognize that a single correction or refutation is unlikely to be enough to help students revise deeply held misconceptions. Instead, guiding students through a process of conceptual change is likely to take time, patience, and creativity.
>
> (Ambrose et al., 2010: 27)

You may need to take a very discussion-intensive approach with in-class worksheets and group work, students working at the board and plenty of repetition of the processes at hand. This type of course can be just as rewarding as those you might find more mathematically interesting; the challenge is pedagogical, not mathematical. The teacher in you will be working overtime, using a few notes the mathematician left behind. While this may not be your dream assignment, you can experience something powerful that you may never have with your more advanced students. If you can take a concept that has haunted a student and help turn on the light, you will experience an addictive exhilaration. When your student gets a piece of self-esteem back, it is priceless for both of you.

The more common experience you will likely have is teaching a course like Precalculus or Calculus, in which the algebra background is supposed to already be in place. While you should be able to assume that most students in your class can do basic processes, like FOIL'ing or factoring, you may also find that some cannot do this well or with ease. It is important to recognize that some students may need this review before they can process the new material you want to teach.

One option for determining how to assess any algebra deficiencies is to consider a prerequisites quiz at the start of the course, as discussed in Chapter 3. You may discover that your class as a whole needs significant algebra instruction or perhaps they come to you very well prepared. A quiz revealing where everyone stands can be useful as you craft the lessons and determine in-class activities. In a class with a weak background, you could open each class with an ungraded quiz on the algebra skills needed for that day, allowing for students to try to recall or rediscover these skills prior to the discussion.

If a quiz on prerequisite skills reveals that you have an even mix of students who are well prepared and those who are not, you have a challenge on your hands! You will need to balance between appeasing students who need a lot of background detail and those who have rock solid prerequisite skills and understandably want you to focus on new material. The first group may be lost and frustrated if you presume too much and the latter will get bored and be less attentive if you presume nothing. Maximizing the learning for everyone can be challenging and it may help to move some review outside of class for those who need additional instruction on prerequisites. Perhaps at the end of each class, you could offer an ungraded homework worksheet that would assist students in practicing the algebra skills needed in the next lesson. If you have already prepared the next lesson, you could include specific algebra problems that will be solved concisely in the next class. For instance, the handout could include a list of quadratics that require factoring within examples you have planned. Students who struggle with this prerequisite skill can work on that piece in advance and focus on the new skills in class.

Regardless of the blend of backgrounds in your class, you may find it useful to offer review exercises in supplemental handouts or online activities and videos. These might be resources provided by your text or available online, such as those through MIT OpenCourseWare and Khan

Academy. Today's student is savvy enough to search these tools out but providing a recommended tool which will mirror your presentation in class may be helpful. Taking this step sends the message that you care about students who need assistance shoring up their foundational skills and reinforces the notion that they can come to you about questions on a prerequisite skill.

Always remember to invite and respond kindly to questions – it takes courage to ask questions, especially on material that is presumed knowledge, and your class may shut down on you if they think you will constantly lecture them on what they should already know how to do. Suppose you are working a problem that involves factoring a quadratic. Before you begin, you might say that it has been awhile since anyone has done this (such as over the summer), so you will go a bit slowly the first couple of times. By acknowledging that students might be rusty, you have opened the door to questions on something students may realize they "should know." Encourage students to check in with you during office hours if they find the algebra to be a struggle during the course of completing the homework.

You may encounter the occasional student whose prerequisite skill level is so low that it will substantially hinder success in your course. If you are encouraging a student to drop to a lower course, emphasize that you want him or her to have the tools that will make your course less frustrating. Point out how much more enjoyable it will be to *succeed* in two math classes as opposed to *failing and repeating* one course due to a lack of background knowledge.

Depending on the level of the institution where you are working, you may even encounter math majors with frightening algebra notions, but generally they should only have a few isolated misconceptions that they have been unable to correct thus far. The upside is that these students are usually aware that they are expected to know these concepts and generally do not fault you for omitting some steps. These students are typically confident enough to ask questions and they should be treated courteously.

Precalculus

Precalculus not only introduces students to functions and limits, but may also include the algebra and trigonometry skills needed in the subsequent calculus course. Sometimes students placed in this course are still struggling with algebra skills and adapting to working with algebraic expressions. You may discover a few calculus-ready students are present, too. Anytime you have a blend of students who have seen little of what you are teaching and those who have seen nearly all of it, it is a challenge. Taking the opportunity to demonstrate how this material relates to similar, more basic concepts will help students who are just learning the material and deepen the understanding for those who have already learned it.

This course lends itself well to drawing parallels to earlier material students may already understand or find less foreign. The two following examples illustrate how you might use familiar concepts to help your class grasp the new ones in your course. Even if you will not be covering these specific skills, you should be able to use this approach for at least some of the concepts in your class.

- Simplifying rational expressions: You can start with an example of reducing a fraction by factoring the numerator and denominator and canceling common factors.

$$\frac{150}{240} = \frac{15 \cdot 10}{24 \cdot 10} = \frac{3 \cdot 5 \cdot 10}{3 \cdot 8 \cdot 10} = \frac{5}{8}$$

Then you can construct a rational expression and demonstrate the similarities, factoring the numerator and denominator, identifying like terms, and canceling. It may help to leave your example with numbers on the board, pointing out parallels, while you simplify the quotient of polynomials.

$$\frac{x^2 - 4}{2x^2 - 3x - 20} = \frac{(x+4)(x-4)}{(2x+5)(x-4)} = \frac{x+4}{2x+5} \quad \text{for } x \neq 4$$

There is a clear difference between these expressions, namely that one is a fixed numerical value while the other involves variables. While obvious, this difference is a vital component of a subtler difference to explore with your class. Note that I have excluded the value of $x = 4$. The last equality above only holds true if we exclude this value of x, since the first two expressions would not exist for $x = 4$ while the last would be 8/13. You can ask the class why we did not have to worry about such a qualification in the earlier example. (While it is also true that we cannot evaluate this expression for $x = -5/2$, this is evident in our final form and all of the equalities stated hold true regardless of input.)

- Polynomial division: You can start with an example of long division with numbers. Your students may not have done this in years, so you may need to remind them how the process works or, better yet, have them try to remind you. Suppose we are interested in using long division to simplify 1232/5. The answer to this division may be reported as:

$$Q = 246, \quad R = 2$$

where Q and R represent quotient and remainder, respectively. Alternatively, the answer could be written in the form of a mixed number. In other words, we could write:

$$\frac{1232}{5} = 246\frac{2}{5}$$

Then you can move on to polynomial long division, showing students how it bears resemblance to the more familiar long division they just completed. Suppose we would like to simplify the following rational expression, using polynomial division.

$$\frac{3x^3 + 7x^2 + x + 4}{x + 2}$$

Pointing back to the long division with numbers, see if your students can correctly set up the long division of the polynomials. As you work through the example, keep returning to what happened with numbers as you discuss what to do next.

We can also discuss the parallels (and difference) in the way we might report answers.

$$Q = 3x^2 + x - 1, \quad R = 6$$

or

$$\frac{3x^3 + 7x^2 + x + 4}{x + 2} = 3x^2 + x - 1 + \frac{6}{x + 2}$$

If you teach synthetic division after students have mastered long division, you can do the same problem with long division first and leave it on the board. After performing the synthetic division, ask students to identify where the same numbers appeared in the long division. Revealing this parallel takes away some of the mystery and demonstrates that this method is *shorthand*, not a trick.

This brings us to the issue of how to handle multiple approaches to material. When two or more methods exist, decide whether it is beneficial to teach multiple methods. On one hand, a student may be completely confused by one approach and find another accessible. On the other hand, seeing more than one solution may be confusing and your students may have trouble differentiating between the two methods. The answer may depend on the level of your students, how similar their backgrounds are, and your available time for the topic.

Consider the following examples of skills which can be approached in at least two, equally valid, ways.

- Simplifying compound rational expressions:
  - One approach is to start by multiplying the top and bottom of the fraction by the least common denominator of all the sub-denominators. This clears all sub-fractions and results in one fraction to simplify. For example:

$$\frac{\frac{1}{x^2}+\frac{x}{2}}{\frac{5}{3}-\frac{x-2}{x}} = \frac{\frac{1}{x^2}+\frac{x}{2}}{\frac{5}{3}-\frac{x-2}{x}} \cdot \frac{2x^2(x-2)}{2x^2(x-2)}$$

$$= \frac{2(x-2)+x^3(x-2)}{10x^2-6x(x-2)} \quad \text{for } x \neq 2$$

$$= \frac{(2+x^3)(x-2)}{4x^2+12x} \quad \text{for } x \neq 2$$

$$= \frac{(2+x^3)(x-2)}{4x(x+3)} \quad \text{for } x \neq 2$$

Note that I have excluded the value of $x = 2$ as discussed in the earlier example. That is, the original expression would not exist for this x-value while the last three would be 0, so the equalities only hold true if we exclude this value of x. (It is also true that we cannot evaluate this expression for the x-values of 0 and -3, but this is evident in our final form and the equalities stated hold true regardless of input.)

  - Another approach is to combine the sub-fractions to achieve one fraction in the numerator and one fraction in the denominator. Once this is accomplished, we multiply by the reciprocal of the simplified denominator fraction.

$$\frac{\dfrac{1}{x^2}+\dfrac{x}{2}}{\dfrac{5}{x-2}-\dfrac{3}{x}} = \frac{\dfrac{2}{2x^2}+\dfrac{x^3}{2x^2}}{\dfrac{5x}{x(x-2)}-\dfrac{3(x-2)}{x(x-2)}}$$

$$= \frac{\dfrac{2+x^3}{2x^2}}{\dfrac{2x+6}{x(x-2)}}$$

$$= \frac{2+x^3}{2x^2} \cdot \frac{x(x-2)}{2x+6} \quad \text{for } x \neq 2$$

$$= \frac{x(2+x^3)(x-2)}{2x^2 \cdot 2(x+3)} \quad \text{for } x \neq 2$$

$$= \frac{(2+x^3)(x-2)}{4x(x+3)} \quad \text{for } x \neq 2$$

If you teach both methods, it may be illustrative to have the class do a couple of problems both ways. This exposes the parallels when you reach similar places in each problem and may help students determine which method they find more appealing.

You may wonder whether the time it takes to adequately cover two approaches is worthwhile. It is a realistic consideration and as with any course topic, you have to weigh its overall value in your course when you determine the time you will devote to it. There can be benefits to your students if they are allowed to choose between the two. Some students will have already seen one of these methods for simplifying the expression and may be confused if only the unfamiliar alternative is presented in class. For students who are new to both, one may feel more natural, possibly because it feels similar to how they were taught to deal with numerical fractions in the past. The risk, as mentioned before, is overwhelming or confusing students with the two approaches.

- Completing the square to solve an equation:
  - One option is to move all constants to the right side of the equation, complete the square on the left side and solve for the variable. For example:

  $$x^2 + 2x - 4 = 0$$

  $$x^2 + 2x = 4$$

  $$x^2 + 2x + 1 = 4 + 1$$

  $$(x+1)^2 = 5$$

  $$x + 1 = \pm\sqrt{5}$$

  $$x = -1 \pm \sqrt{5}$$

  - Alternatively, you might prefer to move everything to the left (if not already there) and complete the square. Then, to solve the equation, begin to move terms to the right. For example:

  $$x^2 + 2x - 4 = 0$$

  $$x^2 + 2x + 1 - 1 - 4 = 0$$

  $$(x+1)^2 - 5 = 0$$

  $$(x+1)^2 = 5$$

  $$x + 1 = \pm\sqrt{5}$$

  $$x = -1 \pm \sqrt{5}$$

In this case, it may depend on how you will use the skill in your course. Problems can arise when students shift from equations to functions. For example, suppose you will ask students to put a quadratic function into standard form by completing the square. Students who learned completing the square by the first approach quickly encounter a problem – the lack of an equal sign! Students invariably try to *solve* the function for x by setting the function equal to zero. You can see from the work below that following the second method results in a seamless transition between functions and equations.

$$f(x) = x^2 + 2x - 4$$
$$f(x) = x^2 + 2x + 1 - 1 - 4$$
$$f(x) = (x+1)^2 - 5$$

Students who learned completing the square by the second approach may still continue their work too far and solve the function set equal to zero, but at least they formulate the correct answer along the way.

Predicting problems such as this is key to a course like Precalculus because students at this level often have a low threshold for even a slight change in methodology and they tend to want to continue performing skills as originally learned. It is impossible the first time teaching a course for you to foresee all possible areas of confusion. There are seemingly small choices you will make along the way that will have unforeseen consequences later, hopefully minor. Paying attention to the nuances of these effects may impact your plans for the next time you teach the course.

If possible, give students leeway to perform the methods that they are most comfortable with, as long as they are demonstrating skills on level with the course. When it really matters and may ultimately otherwise hinder their learning, strongly encourage (or require) the method which is most appropriate. If you do require one method over another equally valid mathematical process, explain your reasoning to the class. Also, make sure that your expectations of the method you will be accepting for credit is clear to students prior to any graded paper.

Opportunities for Opening or Closing Exercises

If you will be drawing parallels to prerequisite material as demonstrated above, you can open class by having students perform the old skill. For instance, you could have students perform the long division with numbers and ask someone put the answer on the board prior to the discussion on polynomial division. The advantage to this approach is that students have actively thought about the long division process, rather than just passively looking on as you remind them. You could close class by asking students to discuss important differences between the earlier skill and the current content. For instance, in the example above we noted that cancelling variable expressions requires attention to the need to potentially exclude certain x-values.

If you have discussed two approaches, you could close class by asking students to compare or distinguish the two. You could ask the class to reflect on why they prefer one method or when one strategy seems more advantageous. Alternatively, you can ask students to discuss why an approach once worked (say when dealing with equations) that no longer applies (such as in the setting of functions.)

Trigonometry Handouts

I often build practice exercises throughout class by introducing a new skill, having the class work one or two problems, then adding content. Students practice a couple more of these more difficult problems and so on. After discussing a sufficient array of increasingly more challenging and varied problems, it can be beneficial to have the class spend 10–15 minutes on a comprehensive mix of the day's or week's content.

The following handouts are examples of this type of mini review. On each, students are asked to demonstrate a number of steps to arrive at a desired trigonometric value without the use of a calculator. Each of the three worksheets serves as a summary exercise of a day or two of classroom discussion. The format of the individual parts closely resembles the format of our classroom examples and that which appears on my tests and quizzes. I formulate the problems in this way to allow students to demonstrate partial knowledge, even if they were unsuccessful in providing accurate final solutions. Replicating the design of the handouts on formal assessments clarifies what information students need to provide on quizzes and exams and familiarizes the context in which they will be tested. The structure also simplifies and expedites my grading since the portions of work that I am intending to grade are given a designated placement on the page.

If you aim to have students complete such a problem without the sort of prompting included here, then you might consider starter exercises such as these, followed by some which eliminate the separate steps. While it is more satisfying to see students perform all the necessary steps of a problem without any prompting, there is a distinction between offering a *hint* and clarifying the information you seek and the work which must be demonstrated.

Copies of the following sheets are available at www.routledge.com/9780367429027.

Finding Trigonometric Values
of Angles between −360° and 360°

1. Label the sides of the common triangles below.

2. In each of the following, draw the location of the original angle on the axes provided, write the reference angle in the blank provided, and circle the sign of the trigonometric value. Compute an exact final answer for each.

a) $\cos(150°) = \pm \cos$ _____ = _____
   Circle sign   *reference angle*   *final answer*

b) $\sin(330°) = \pm \sin$ _____ = _____
                  reference angle   *final answer*

c) $\csc(-60°) = \pm \csc$ _____ = _____
                  reference angle   *final answer*

d) $\cos(240°) = \pm \cos$ _____ = _____
                  reference angle   *final answer*

e) $\cot(-135°) = \pm \cot$ _____ = _____
                  reference angle   *final answer*

Figure 5.1 Finding Trigonometric Values of Angles Between (−360°, 360°)

Finding Trigonometric Values of Angles Outside $(-360°, 360°)$

1. Label the sides of the common triangles below.

2. In each of the following, draw the location of the original angle on the axes provided and find a representation of the angle's position that falls between 0° and 360°. Write the reference angle in the blank provided and circle the sign of the trigonometric value. Compute an exact final answer for each.

a) $\sin(1680°) = \sin$ _____ $= \pm \sin$ _____ $=$ _____
                              angle between 0° and 360°   reference angle   final answer
                                             Circle sign

b) $\csc(-750°) = \csc$ _____ $= \pm \csc$ _____ $=$ _____
                              angle between 0° and 360°   reference angle   final answer

c) $\cos(405°) = \cos$ _____ $= \pm \cos$ _____ $=$ _____
                           angle between 0° and 360°   reference angle   final answer

d) $\cot(-450°) = \cot$ _____ $= \pm \cot$ _____ $=$ _____
                            angle between 0° and 360°   reference angle   final answer

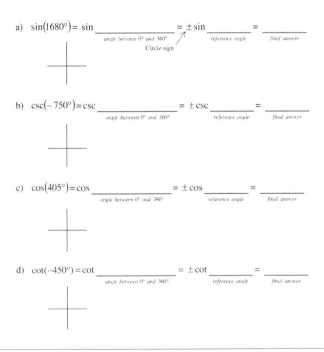

Figure 5.2 Finding Trigonometric Values of Angles Outside $(-360°, 360°)$

Finding Trigonometric Values
Using the Unit Circle

1. Label each of the 5 points marked on the first quadrant of the unit circle pictured below by stating the radian angle and the (x, y) point on the unit circle.

2. For each of the following, draw the given angle on the unit circle, write the reference angle in the blank provided, and circle the sign of the trigonometric value. Compute an exact final answer for each.

   a) $\sin\left(\dfrac{3\pi}{4}\right) = \pm \sin$ _____ = _____

   Circle sign → reference angle (in radians or °) final answer

   b) $\cos\left(\dfrac{4\pi}{3}\right) = \pm \cos$ _____ = _____

   reference angle (in radians or °) final answer

   c) $\cot\left(\dfrac{7\pi}{6}\right) = \pm \cot$ _____ = _____

   reference angle (in radians or °) final answer

   d) $\cos(3\pi) = \pm \cos$ _____ = _____

   reference angle (in radians or °) final answer

   e) $\sin\left(\dfrac{17\pi}{6}\right) = \pm \sin$ _____ = _____

   reference angle (in radians or °) final answer

Figure 5.3 Finding Trigonometric Values Using the Unit Circle

Calculus I: Differential Calculus

In differential calculus, you may encounter a mix of students who have had a year of calculus in high school and those who have only had precalculus. It can be difficult when you have overly prepared students in the class but teach the course as though it is all new material. When you review precalculus topics, such as equations of lines, you can assume some knowledge but make sure to remind students of the basics. Always remember that math is easy to forget when left unpracticed and few students are thinking about it over the summer.

A point that must be emphasized in differential calculus is that students need to learn the basic differentiation rules well and relatively quickly. What you cover in one or two days of class will be used *every day* for the rest of the semester. If a student's mastery of the basic skills is insufficient, his or her flawed techniques will greatly hinder attempts to master the product, quotient, and chain rules. Alerting your students to the importance of this material and encouraging students to stop by office hours with questions is great, but consider offering ample opportunities for retrieval and practice of these rules in class.

You might find you are able to discuss a variety of examples on topics such as the product or quotient rule with ample time remaining in class. This is a good example of when the class composition could fool you into moving too quickly. Students who have had calculus before will recognize the rule and may recall having done fairly well with it in the past. Those students may answer your questions quickly or become restless, leading you to believe you have belabored the coverage. Most students will require quite a bit of practice before these skills are mastered, so these offer a perfect opportunity for in-class exercises. Even students who have learned the material previously will benefit from practice, so this is time well spent prior to more complicated examples involving multiple rules. Additionally, because some students will know answers in advance, it is important to give the class time to consider your questions before taking proposed solutions.

In an advanced differential calculus course, a particular challenge you may face is having students create and *comprehend* epsilon-delta proofs for limits. Students will generally get the hang of the proof process after a while, but may still struggle with what is happening conceptually. Starting with several examples for which you can provide graphical demonstrations of the behavior may be helpful and your text may offer interactive exercises or videos to assist in this. Conceptual discussions and representations aid your students' ability to construct proofs correctly and to avoid simply memorizing a format.

Curve-sketching provides a wonderful summary of derivative applications in this course but can present challenges for testing. Students often find satisfaction in using so much of what they have learned over the past month or two to create a graph. Naturally, the graphs are more interesting and challenging for functions that require more complicated differentiation, but this poses a difficulty when it comes to an exam. If a student makes even a relatively minor error early, the entire problem may become overly messy or trivialized. Alternatively, your students may get snarled in the derivative and never reach a point where they are able to demonstrate their knowledge within the application.

There are several options to lessen the grading hassles these lengthy problems can produce. You could avoid complicated functions for this application, even selecting polynomials with coefficients chosen so that the critical and inflection points do not contain messy values. For more complicated functions, such as those involving rational or transcendental components, you could provide the first and second derivatives. Alternatively, you could ask students to draw a graph, provided a list of information about an unspecified function and its derivatives. All of these suggestions essentially suggest that you consider testing differentiation skills separately from graphing applications. While you will need to test your students' ability to take derivatives of a variety of functions, blending intricate differentiation with a lengthy application on an examination may mean that you are not able to accurately assess their abilities in either.

Opportunities for Derivation or Proof

You probably do not have time to prove every theorem and formula you encounter, but encouraging your students to consider and discuss the crux of most develops their critical thinking and appreciation of required hypotheses. The Mean Value Theorem and Rolle's Theorem are very accessible, at least conceptually, to many students.

Proving derivative formulas for a host of functions serves to help students understand why the formulas are correct and provide practice with the definition of the derivative or the derivative rules (e.g. using the quotient rule to find the derivative of $y = \tan x$). Think through whether students have all of the necessary skills to perform a derivation. For instance, to prove the derivative of $y = \sin x$, you'll need to use the angle sum formula and know the values of the limits $\lim_{\Delta x \to 0} \dfrac{\cos \Delta x - 1}{\Delta x}$ and $\lim_{\Delta x \to 0} \dfrac{\sin \Delta x}{\Delta x}$. Perhaps these are second nature to you, but if you have not discussed these in your class, do not assume that your students have seen or remember them.

It can be easy to overwhelm students with material that looks pretty straightforward to you. If you are unsure what level of derivation or proof is expected in the course you are teaching, ask a colleague or superior for guidance. If you have some latitude, then test the waters with short proofs and see how well your students are able to engage with the material. Can they offer suggestions? If broken into groups, can they make headway? How do they respond when the solution is ultimately revealed?

If you are teaching a low-level calculus course not intended for majors, students may get lost in lengthy formal equations required for some proofs, but consider discussing the intuition behind theorems. Stating theorems as facts we just accept does nothing to further mathematical thinking and limits the comprehension and retention of the theorems at hand.

Opportunities for Opening and Closing Exercises

Since proficiency with differentiation rules is so vital to success in this course, designating time to recall and use these rules is one option for the start and end of class. You might use a closing quiz to ask students to recall rules learned in that session. After students have practiced in the homework, you could open the following class with a similar or possibly more challenging retrieval exercise. Another exercise is asking students to identify the rule(s) that would be required to differentiate a given function.

Opening and closing quizzes can be used to further drive home the conceptual aspects of limit proofs. For instance, you could ask students to describe what ε and δ represent and which one you can control. Alternatively, you might ask why the ability to achieve an ε–distance by controlling a δ–distance proves the limit holds. Asking them to discuss this generally, rather than in terms of a specific example, requires they understand the concept. Without this level of comprehension, students may only be memorizing steps to determine a sufficient δ.

Graphing Techniques Handouts

Here, I will examine two possible styles of handouts which cover the same material. I created the first handout years ago as a review of graphing

Not Recommended:
Finding Relative Extrema, Monotonicity, and Concavity
For Polynomials and Rational Functions

Relative Extrema and Critical Numbers
- First note the domain of your function.
- Solve $f'(x) = 0$ for x and find any x-values for which $f'(x)$ is undefined. The solutions which are **in the domain** of $f(x)$ are the <u>critical numbers</u>.
- Perform either the first or second derivative test:
    First derivative test:
        Make a sign chart for $f'(x)$, using the critical numbers and discontinuities of $f(x)$.
        If the sign of $f'(x)$ changes from $+$ to $-$, you have found a relative max (unless the x-value is not in the domain of $f(x)$).
        If the sign of $f'(x)$ changes from $-$ to $+$, you have found a relative min (unless the x-value is not in the domain of $f(x)$).
    Second derivative test:
        Plug each critical number, $x = c$, into $f''(x)$.
        If $f''(c) > 0$, then you have a relative min at $x = c$.
        If $f''(c) < 0$, then you have a relative max at $x = c$.
        If $f''(c) = 0$ or does not exist, then the second derivative test is inconclusive; use the first derivative test to find extrema.
- Find the y-values of the extrema by plugging the x-values into the <u>original function</u>, $f(x)$. State each final answer as an (x, y) point.

Determining Intervals Where a Function Increases/Decreases
- First note the domain of your function.
- Solve $f'(x) = 0$ for x and find any x-values for which $f'(x)$ is undefined. The solutions which are **in the domain** of $f(x)$ are your <u>critical numbers</u>.
- Make a sign chart for $f'(x)$, using the critical numbers and all discontinuities of $f(x)$.
    $f(x)$ increases on the intervals where $f'(x) > 0$.
    $f(x)$ decreases on the intervals where $f'(x) < 0$.
- Give one answer for increasing (connecting intervals together with union symbols) and one answer for decreasing (connecting intervals together with union symbols).

Determining Inflection Points and Intervals Where a Function is CCU/CCD
- First note the domain of your function.
- Solve $f''(x) = 0$ for x and find any x-values for which $f''(x)$ is undefined. The solutions which are **in the domain** of $f(x)$ are your <u>possible inflection points</u>.
- Make a sign chart for $f''(x)$, using the possible inflection points and the discontinuities of $f(x)$.
    $f(x)$ is concave up (CCU) on the intervals where $f''(x) > 0$.
    $f(x)$ is concave down (CCD) on the intervals where $f''(x) < 0$.
- Give one answer for CCU (connecting intervals together with union symbols) and one answer for CCD (connecting intervals together with union symbols).
- Inflection points occur only for x-values in the domain of $f(x)$ where concavity changes (i.e. where the sign changes on your sign chart for $f''(x)$). Find the y-values of the inflection points by plugging the x-values into the <u>original function</u>, $f(x)$. State each final answer as an (x, y) point.

Figure 5.4 Example of a review sheet lacking engagement. Finding Relative Extrema, Monotonicity and Concavity

techniques for my calculus course, but did not continue to distribute it because students did not seem to utilize it. The revised handouts offer a vast improvement upon my strategy for having students review and engage with the material.

So, why didn't my students find the first sheet to be a useful resource? It provided all the details they needed to complete the relevant problems. Isn't that what students want from us – one piece of paper that lists everything they should know? Unfortunately, this sheet may have only been helpful to the strongest students who were ready to nail down the fine details of curve sketching. They would be able to read over the contents and catch the little bits they did not know or had misunderstood. Students who were still trying to learn and remember the basics may have just seen a mass of excessive information. This probably only served to discourage them.

In addition to overwhelming students, a danger in using such a detailed sheet is that students may believe they should just *memorize* the information printed here. Unfortunately, memorization does not often lead to much success with math at this level. Using a review sheet which just tells students everything is similar to lecturing all period with a disregard for student engagement.

In the revised subsequent handouts, students are asked to actively think about the critical portions of each statement and fill in the information themselves. Not only does it guide the student through the mathematical process but it also seeks to develop the thought process behind it. It will feel far more approachable to weaker students and remain useful to all student levels.

There is also more thought put into the separation of concepts. The distinct process of the second derivative test is moved to an entirely separate handout to further the point that, in general, one would not perform both the first and second derivative tests. Concavity is also handled separately to distinguish it from the extrema. Note how one intense handout morphed into three when I considered the student experience.

Each of the improved handouts could be used as a summary review shortly after learning the skills or as part of an exam review. Any individual problem on these sheets could be used for an opening or closing quiz. Revisiting these questions throughout the progression of these topics will serve to clarify and imbed these topics in your students' minds. Regardless of how they are used, the process of filling in the missing information, rather than reading it, will lead to improved learning and retention of the material (Brown, Roediger, & McDaniel, 2014: 208).

Finding Relative Extrema and Intervals of Monotonicity
For Polynomials and Rational Functions

First Derivative Test Method

1. Determine the domain of $f(x)$.

    How? Determine the x-values where $f(x)$ does not exist. The domain will consist of all real numbers, excluding these x-values.

    Why? The excluded x-values will not signify possible extrema of polynomials or rational functions, but monotonicity could change at these x-values.

2. Find all critical numbers for $f(x)$.

    How? Find x-values in the domain of $f(x)$ where $f'(x) = 0$ or $f'(x)$ does not exist.

    Why? These are the x-values where $f(x)$ may have relative extrema or where monotonicity could change.

3. Perform the first derivative test:
    - Make a sign chart for ___$f'(x)$___ using the ___critical numbers___ and ___discontinuities of $f(x)$___.

    - When the sign of ___$f'(x)$___ changes from + to − at one of the x-values on the chart, this indicates the occurrence of a relative ___max___ of ___$f(x)$___ unless the x-value is ___not in the domain of $f(x)$___.

    - Where the sign of ___$f'(x)$___ changes from − to +, at one of the x-values on the chart, this indicates the occurrence of a relative ___min___ of ___$f(x)$___ unless the x-value is ___not in the domain of $f(x)$___.

    - If the sign of ___$f'(x)$___ does not change at an x-value on the sign chart, this indicates there is ___no relative extrema___ at that x-value.

4. Summarize results.
    - State the extrema.
      Find the y-values of the extrema by plugging each ___$x =$ critical number___ into ___$f(x)$___. State each final answer as ___an (x, y) point___ and label each as a relative ___max___ or relative ___min___.

    - State the intervals of monotonicity.
      $f(x)$ increases on the intervals where ___$f'(x) > 0$___.

      $f(x)$ decreases on the intervals where ___$f'(x) < 0$___.

      Use ___interval___ notation to report the answers, connecting multiple intervals together with ___union symbols___.

Work out an example on the back or a new sheet of paper.

Figure 5.5 Finding Relative Extrema and Intervals of Monotonicity for Polynomials and Rational Functions

Finding Relative Extrema
For Polynomials and Rational Functions

Second Derivative Test Method

1. Determine the domain of $f(x)$.

    How? *Determine the x-values where $f(x)$ does not exist. The domain will consist of all real numbers, excluding these x-values.*

    Why? *The excluded x-values will not be extrema of polynomials or rational functions.*

2. Find all critical numbers for $f(x)$.

    How? *Find x-values in the domain of $f(x)$ where $f'(x) = 0$ or $f'(x)$ does not exist.*

    Why? *These are the x-values where $f(x)$ may have relative extrema.*

3. Perform the second derivative test:

    - Plug each __critical number, $x = c$__ into __$f''(x)$__ .
    - If __$f''(c)$__ > 0, then there is a relative __min__ of __$f(x)$__ at $x = $ __c__ .
    - If __$f''(c)$__ < 0, then there is a relative __max__ of __$f(x)$__ at $x = $ __c__ .
    - If __$f''(c)$__ = 0 or does not exist, then the second derivative test is __inconclusive__ and we will determine the extrema by using the __first derivative test__ instead.

4. State the extrema.

    Find the y-values of the extrema by plugging each __$x = c$__ into __$f(x)$__ . State the final answer as __an (x, y) point__ and label each as a relative __max__ or relative __min__ .

Work out an example below or on a new sheet of paper.

Figure 5.6 Finding Relative Extrema for Polynomials and Rational Functions

Determining Inflection Points and Concavity
For Polynomials and Rational Functions

1. Determine the domain of $f(x)$.

    How? *Determine the x-values where $f(x)$ does not exist. The domain will consist of all real numbers, excluding these x-values.*

    Why? *The excluded x-values will not be inflection points, but concavity could change at these x-values.*

2. Find all possible inflection points for $f(x)$.

    How? *Determine the x-values in the domain of $f(x)$ for which $f''(x) = 0$ or $f''(x)$ does not exist.*

    Why? *These are the x-values where $f(x)$ may have inflection points, i.e. where concavity could change.*

3. Make a sign chart.
    - Make a sign chart for __$f''(x)$__ using the __possible inflection points__ and __discontinuities of $f(x)$__.

    - Where the sign of __$f''(x)$__ changes from + to −, $f(x)$ has __an inflection point__ unless the x-value is __not in the domain of $f(x)$__.

    - Where the sign of __$f''(x)$__ changes from − to +, $f(x)$ has __an inflection point__ unless the x-value is __not in the domain of $f(x)$__.

    - If the sign of __$f''(x)$__ does not change at an x-value on the sign chart, there is __no inflection point__ for $f(x)$ at that x-value.

4. Summarize results.
    - State each inflection point.
      Find the y-value of each inflection point by plugging __the x-value__ into __$f(x)$__. State each final answer as __an (x, y) point__ and label each as an inflection point.

    - State the intervals of concavity.
      $f(x)$ is __concave up (CCU)__ on the intervals where __$f''(x) > 0$__.
      $f(x)$ is __concave down (CCD)__ on the intervals where __$f''(x) < 0$__.

    - Use __interval__ notation to report the answers, connecting multiple intervals together with __union symbols__.

    Work out an example on the back or on a new sheet of paper.

Figure 5.7 Determining Inflection Points and Concavity for Polynomials and Rational Functions

Notes:

- On the revised handouts, suggested answers are given in italics. Blank versions, along with these keys, are available at www.routledge.com/9780367429027.
- The handouts describe methods for polynomials and rational functions. Piecewise functions would need additional attention. Specifically, extrema could occur at a discontinuity.

Calculus II: Integral Calculus

In integral calculus, students are faced with a wide variety of integration techniques. For this reason, even students who have had high school calculus often see a decent amount of new material and benefit from revisiting what they have learned. The class may be a little overwhelmed once a sufficient number of techniques have been introduced. To address this problem, consider a cumulative approach to the material in which you persistently return to earlier skills.

Students can often master the individual integration tools but are challenged when a mix of problems are put before them. It is not uncommon to see students keep reaching for the last tool which sits atop their mathematical toolbox. It is fresh in their minds and they are eager to grab it and demonstrate its use, even if it is a high-powered drill when a rudimentary mallet is sufficient. Encourage your students to identify certain integrands which should cry out for a method (such as xe^x), so that they look for recognizable forms before blindly going through the list of available methods.

Learning how to sort through all of the skills they have learned takes a bit of time but interleaving material and using the cumulative quizzes can help. You can interleave the integration techniques, as discussed in *A Typical Class* and by repeatedly circling back to earlier methods in class and on quizzes. Regularly including a cumulative portion of quizzes discussed in the introduction of Chapter 4 offers a perfect opportunity to gradually add to your students' integral toolbox. Consider including at least one old integration method on each quiz testing new methods in an effort to keep the knowledge continuous and cumulative instead of disjoint. Selecting similar integrands to those which required a new approach may prove especially useful in reminding students of when that skill was not necessary. For instance, if your quiz involved an integral involving the form xe^x, you might also include an integral of the form Ce^x.

Graphical representations are vital when covering volumes of solids of revolution, allowing you to carefully address how the solid is sliced into disks and washers or built with shells and how this relates to the differential. If you are artistically challenged, you may want to use overheads, computer projections or physical models in place of drawing on the board but recognize that students will be attempting to draw such diagrams in their notes, assignments, quizzes, and tests. Unless you plan to always provide students with a completed diagram, you should demonstrate how these can be roughly constructed by hand.

Since it is vital to have the correct graph prior to setting up integrals for area or volume, you may want to give a pretest on the graph shapes your students know. Some may have learned or retained little knowledge of common graph shapes and might need some assistance assembling a library. Consider a brief review of the graphs you expect students to know and how to use the equations for quick rough sketches rather than plotting points. Regardless of the level of review you offer, your class will appreciate clear guidelines of which functions they should know how to sketch for application problems.

Opportunities for Derivation or Proof

There is a wide variety of calculus courses and some may not delve into derivation often, but there are opportunities for all levels. For instance, the derivation of the formula for integration by parts is interesting to a spectrum of audiences. True, it is fairly simple, but this makes it accessible for many ability levels and can be used as an in-class group exercise or independent assignment depending on the course. Additionally, the use of product rule ties in older material and extends its purpose beyond basic calculation of function derivatives. The Fundamental Theorem of Calculus proven carefully is time consuming, but also illustrative to the proper audience. If you do not feel that detailed proofs are appropriate for the course or audience at hand, consider what aspects you could discuss to better illuminate why formulas work.

Opportunities for Opening or Closing Exercises

A quick, but impactful, exercise is to ask students to explain choices regarding methods of integration. They can identify the proper strategy for a given integral and explain what properties of the integrand suggested this approach.

Another option is to ask them to distinguish between similar integrands, such as $\sin x$, $x \sin x^2$, and $x \sin x$.

In area and volume applications, consider employing strategies which will encourage your students to avoid blind memorization as to when to integrate with respect to x or y. Creating opening or closing questions, perhaps based solely on a diagram, which ask for descriptions of technique and justification of the chosen differential, gets at the *why* as opposed to the *how*. "How" is often approached by memorization but "why" encourages reflection.

Similarly, consider having your students discuss why the method of cylindrical shells can be beneficial and when it is preferable over the disk or washer approach. You could open or close class with an ungraded quiz asking students to determine which method to apply to find the volume of a solid of revolution and to explain *why* they made that choice. Propose an example solid which requires two integrals when done by the washer method but only one with cylindrical shells, such as the volume of the solid generated by revolving the region bounded by $y = x^2$, $y = 0$, $x = 1$, and $x = 3$ around the y-axis.

Integration Strategy Handouts

Here we have another example in which my initial attempt to provide a review sheet was unsuccessful for the masses. I created the original handout in response to a talented and dedicated group of students feeling overwhelmed by all the techniques they needed to consider on an upcoming comprehensive final examination. I spent a good amount of time crafting the hierarchy and examples to provide as concise, but thorough, an overview as I could.

As I handed out the review sheet, I was disappointed to see that students were not in fact relieved and appreciative. Their anxiety only seemed to increase. At the time, I thought it was only the initial impact of seeing the detailed sheet, but assumed their perspective would change as they read through it later and realized that they had a good grasp of the material it summarized. This may have been true for some but it is possible other students simply put it aside, feeling there were too many details to absorb.

If such a detailed review appeals to you, I would encourage you to have the class create the review sheet together. To structure the review, you could use fill-in worksheets, such as the revised sheets provided at the end of this section. If desired, students could subsequently produce their own compacted review guide. Alternatively, you can use handouts such as these as you go through the material by completing each subsection shortly after you have

Not Recommended:

<ins>Integration Strategy Comprehensive Review</ins>

1. **Simplify** the integrand when possible.
   - <ins>Products</ins>: If the product is one that is easily multiplied out, this may be the simplest (or only) way to solve the integral. For example:

   $$\int x^2(2-x)dx = \int (2x^2 - x^3)dx = \frac{2}{3}x^3 - \frac{1}{4}x^4 + C$$

   - <ins>Fractions</ins>: If the denominator is one term, you can break the fraction apart, writing each piece of the numerator over the denominator. For example:

   $$\int \frac{1-x^5}{6x^2}dx = \int \left(\frac{1}{6x^2} - \frac{x^5}{6x^2}\right)dx = \int \left(\frac{1}{6}x^{-2} - \frac{1}{6}x^3\right)dx$$
   $$= -\frac{1}{6}x^{-1} - \frac{1}{24}x^4 + C = -\frac{1}{6x} - \frac{1}{24}x^4 + C$$

2. Look for a basic ***u*-substitution**.
   - <ins>Powers</ins>: If you have an expression raised to a power, try a *u*-substitution for the expression. For example, letting $u = 6 - x$ in the following:

   $$\int (6-x)^{10}dx = \int -u^{10}du = -\frac{1}{11}u^{11} + C = -\frac{1}{11}(6-x)^{11} + C$$

   - <ins>Trigonometric Functions</ins>:
     - Try a *u*-substitution for the angle. For example, letting $u = 6x$ in the following:

     $$\int \sin(6x)\, dx = \int \frac{1}{6}\sin(u)\, du = -\frac{1}{6}\cos(u) + C = -\frac{1}{6}\cos(6x) + C$$

     - Try a *u*-substitution for an entire trig function. For example, letting $u = \sin(3x)$ in the following:

     $$\int \sin(3x)\cos(3x)\, dx = \int \frac{1}{3}u\, du = \frac{1}{6}u^2 + C = \frac{1}{6}\sin^2(3x) + C$$

     Note: For the above example, you could choose $u = \sin(3x)$ or $u = \cos(3x)$. If one of the functions in the example above is raised to a power, then *u* will be that function. So, in the following example, we purposefully choose $u = \sin(3x)$. For more complicated problems, see #4.

     $$\int \sin^2(3x)\cos(3x)\, dx = \int \frac{1}{3}u^2\, du = \frac{1}{9}u^3 + C = \frac{1}{9}\sin^3(3x) + C$$

     - Try rewriting functions in terms of sine and/or cosine. (The *u*-substitution $u = \cos x$ is also used below.)

     $$\int \tan x\, dx = \int \frac{\sin x}{\cos x}dx = \int -\frac{1}{u}du = -\ln|u| + C = -\ln|\cos x| + C$$
     (Equivalently, this answer can be written as: $\ln|\sec x| + C$.)

Figure 5.8 Example of providing excessive detail in a review sheet. Integration Strategy Comprehensive Review

- Exponentials:
  - Try a u-substitution for the exponent. For example, let $u = 3x + 1$ in the following:

  $$\int e^{3x+1} dx = \int \frac{1}{3} e^u du = \frac{1}{3} e^u + C = \frac{1}{3} e^{3x+1} + C$$

  Or let $u = 2x^2 - 4$ in the following:

  $$\int x 7^{2x^2-4} dx = \int \frac{1}{4} 7^u du = \frac{1}{4\ln 7} 7^u + C = \frac{1}{4\ln 7} 7^{2x^2-4} + C$$

  - Try a u-substitution that includes the entire exponential. For example, let $u = e^x + 1$ in the following:

  $$\int \frac{e^x}{e^x + 1} dx = \int \frac{1}{u} du = \ln|u| + C = \ln|e^x + 1| + C = \ln(e^x + 1) + C$$

- Logarithms: Try a u-substitution for the entire log term. For example, let $u = \ln x$ in the following:

  $$\int \frac{\ln x}{x} dx = \int u \, du = \frac{1}{2} u^2 + C = \frac{1}{2} (\ln x)^2 + C$$

- Products: If the product cannot be multiplied out, or that would require significant work, then try a u-substitution. For example, let $u = 2 - x^2$ in the following:

  $$\int x\sqrt{2 - x^2} \, dx = \int -\frac{1}{2} u^{1/2} du = -\frac{1}{3} u^{3/2} + C = -\frac{1}{3}(2 - x^2)^{3/2} + C$$
  $$= -\frac{1}{3}\sqrt{(2 - x^2)^3} + C$$

- Fractions: If the fraction cannot be simplified as discussed in #1, look for the following substitutions as a possible way to solve the integral.
  - If there is an expression raised to a power, try a u-substitution for the expression. For example, let $u = 4 - x$ in the following:

  $$\int \frac{2}{(4 - x)^{10}} dx = \int -2u^{-10} du = \frac{2}{9} u^{-9} + C = \frac{2}{9}(4 - x)^{-9} + C$$
  $$= \frac{2}{9(4 - x)^9} + C$$

  - If it appears that the numerator may be related to the derivative of the denominator, try letting $u =$ denominator. For example, let $u = x^5 - 1$ in the following:

  $$\int \frac{3x^4}{x^5 - 1} dx = \int \frac{3}{5} \frac{1}{u} du = \frac{3}{5} \ln|u| + C = \frac{3}{5} \ln|x^5 - 1| + C$$

  - If the denominator factors, consider partial fractions (see #3).

Figure 5.8 Continued

3. If the integrand is a proper rational function (i.e. a fraction of polynomials with the degree of the numerator less than the degree of the denominator), try simplifying the integrand using **partial fractions**. Factor the denominator and follow the guidelines below to construct the decomposition.

- Linear factors which are not repeated will have a constant in the numerator.

$$\frac{x}{(x-1)(x+2)} = \frac{A}{x-1} + \frac{B}{x+2}$$

- Linear repeated factors will have a constant in the numerator and each power up to the power of the repeated factor must be represented in the denominators.

$$\frac{x}{(x-1)(x+2)^3} = \frac{A}{x-1} + \frac{B}{x+2} + \frac{C}{(x+2)^2} + \frac{D}{(x+2)^3}$$

- Irreducible quadratic factors which are not repeated will have linear numerators.

$$\frac{x}{(x^2+1)(x+2)} = \frac{Ax+B}{x^2+1} + \frac{C}{x+2}$$

- Irreducible quadratic repeated factors will have linear numerators and each power of the factor must be represented in the denominators.

$$\frac{x}{(x^2+1)^2(x+2)} = \frac{Ax+B}{x^2+1} + \frac{Cx+D}{(x^2+1)^2} + \frac{E}{x+2}$$

Notes:
- You will use $\int \frac{1}{u} du = \ln|u| + C$ and $\int \frac{1}{x^2+a^2} dx = \frac{1}{a}\tan^{-1}\left(\frac{x}{a}\right) + C$ for many of these problems.

- These integrals will often have answers involving the natural log. Use log rules to combine terms.

4. If the integrand consists of **powers of trig functions**, try the following substitutions.

- Integrals with sine and cosine:
    - If the power on either sine or cosine is odd, pull off one copy of that trig function and convert the rest to the other trig function, using $\sin^2 x + \cos^2 x = 1$. If both powers are odd, you can pick either one to convert.
    - If both are raised to even powers, convert both using $\sin^2 x = \frac{1}{2}(1 - \cos(2x))$ and $\cos^2 x = \frac{1}{2}(1 + \cos(2x))$.

- Integrals with tangent and secant:
    - If the power on secant is even, pull off one copy of $\sec^2 x$, and convert the rest of the secants to tangents, using $\sec^2 x = 1 + \tan^2 x$.
    - If the power on tangent is odd, pull off one copy of secant and one copy of tangent, and convert the rest of the tangents to secants, using $\tan^2 x = \sec^2 x - 1$.
    - If neither is true, experiment.

Figure 5.8 Continued

- Integrals with cotangent and cosecant:
    * If the power on cosecant is even, pull off one copy of $\csc^2 x$, and convert the rest of the cosecants to cotangents, using $\csc^2 x = 1 + \cot^2 x$.
    * If the power on cotangent is odd, pull off one copy of cosecant and one copy of cotangent, and convert the rest of the cotangents to cosecants, using $\cot^2 x = \csc^2 x - 1$.
    * If neither is true, experiment.

5. If the integrand involves **square roots** and the standard u-substitution fails, consider the following trigonometric substitutions.

    * $\sqrt{a^2 - x^2}$: $x = a \sin\theta$, $-\frac{\pi}{2} \leq \theta \leq \frac{\pi}{2}$
    * $\sqrt{a^2 + x^2}$: $x = a \tan\theta$, $-\frac{\pi}{2} < \theta < \frac{\pi}{2}$
    * $\sqrt{x^2 - a^2}$: $x = a \sec\theta$, $0 \leq \theta < \frac{\pi}{2}$ or $\pi \leq \theta < \frac{3\pi}{2}$

6. Try **integration by parts**: $\int u\, dv = uv - \int v\, du$ or $\int_a^b u\, dv = uv\Big|_a^b - \int_a^b v\, du$

    * Recognize integral forms which are solved by integration by parts, such as:

    $$\int x^n e^x dx, \int x^n \sin x\, dx, \int x^n \cos x\, dx, \int x^n \ln x\, dx, \int \ln x\, dx$$

    * Remember the special approach required for an integral such as $\int e^x \sin x\, dx$.

Figure 5.8 Continued

completed the material it covers. For exam review, students can fill in new blank sheets. Consider following this exercise with a worksheet containing a scramble of the types of integrals they could see on the exam and ask them only to identify which method of integration they would utilize. While they might use the list for additional practice of skills later, having students first try to discern the best method encourages mathematical thinking and better prepares them to properly strategize during an exam.

On the revised review sheets, students create their own examples. This process helps them think about the role each piece of the integrand is playing. It also helps them consider the overall process instead of how to attack a particular example you put before them. Remind students not to look up answers or examples until after they have completed the sheet.

The original handout was long, but when it is expanded into an interactive set of sheets, we can genuinely appreciate how much information is present. This reveals exactly why the original was overwhelming to my students. There is a lot of information there and each piece needs to be given its own moment in the sun. The first handout also potentially promotes the ill-advised approach of memorizing this information, whereas the second

Integration Strategy
A First Swing at Products, Fractions, and Powers

1. The integrand is a **product of polynomials**:

   Simplify the integrand by __multiplying out__ a product of polynomials and use the __power rule__ to find the antiderivative.

   Work out an example:

   $$\int x^2(2-x)dx = \int (2x^2 - x^3)dx = \frac{2}{3}x^3 - \frac{1}{4}x^4 + C$$

2. The integrand is a **rational function** (i.e. a fraction of __polynomials__):

   - If the denominator is of the form __Cx^n__ and the numerator has more than one term, simplify the integrand by __breaking the rational function into separate fractions__. Rewrite each term in the form __Dx^k__ and use the __power rule__ to find the antiderivative.

     Work out an example:

     $$\int \frac{1-x^5}{6x^2}dx = \int \left(\frac{1}{6x^2} - \frac{x^5}{6x^2}\right)dx = \int \left(6x^{-2} - \frac{1}{6}x^3\right)dx$$
     $$= -\frac{1}{6}x^{-1} - \frac{1}{24}x^4 + C = -\frac{1}{6x} - \frac{1}{24}x^4 + C$$

   - If the denominator is a factor of the numerator, __cancel the denominator__ and use the __power rule__ to find the antiderivative.

     Work out an example:

     $$\int \frac{1-x^2}{x+1}dx = \int \left(\frac{(1-x)(1+x)}{x+1}\right)dx = \int (1-x)dx \text{ for } x \neq -1$$
     $$= x - \frac{1}{2}x^2 + C \text{ for } x \neq -1$$

3. The integrand involves an **expression raised to a power**, try a __u-substitution__ for the __expression__.

   Work out an example where the expression is a polynomial:

   $$\int (6-x)^{10}dx \quad \begin{matrix} u = 6-x \\ du = -dx \end{matrix}$$
   $$= \int -u^{10}du = -\frac{1}{11}u^{11} + C = -\frac{1}{11}(6-x)^{11} + C$$

   Work out an example in which the expression is not a polynomial:

   $$\int \sqrt{2-x}\, dx = \int (2-x)^{\frac{1}{2}}dx \quad \begin{matrix} u = 2-x \\ du = -dx \end{matrix}$$
   $$= \int -u^{\frac{1}{2}}du = -\frac{2}{3}u^{\frac{3}{2}} + C = -\frac{2}{3}(2-x)^{\frac{3}{2}} + C = -\frac{2}{3}\sqrt{(2-x)^3} + C$$

Figure 5.9 Integration Strategy: A First Swing at Products, Fractions, and Powers

Integration Strategy
A Second Swing at Products

When the integrand is a **product which cannot be multiplied out** or cannot be easily multiplied out:

1. Try a __u-substitution__ for the expression which is problematic to multiply out.

   Work out an example in which the expression would be laborious to multiply out.

   $$\int x(2-x^2)^{99}\, dx \quad \begin{matrix} u = 2 - x^2 \\ du = -2x\, dx \end{matrix}$$

   $$= \int -\frac{1}{2} u^{99}\, du = -\frac{1}{200} u^{100} + C = -\frac{1}{200}(2-x^2)^{100} + C$$

   Work out an example in which the expression could not be multiplied out.

   $$\int x\sqrt{2-x^2}\, dx \quad \begin{matrix} u = 2 - x^2 \\ du = -2x\, dx \end{matrix}$$

   $$= \int -\frac{1}{2} u^{1/2}\, du = -\frac{1}{3} u^{3/2} + C = -\frac{1}{3}(2-x^2)^{3/2} + C$$
   $$= -\frac{1}{3}\sqrt{(2-x^2)^3} + C$$

2. Try __integration by parts__, for which we have the indefinite and definite integral formulas:

   $$\int u\, dv = uv - \int v\, du \quad \text{and} \quad \int_a^b u\, dv = uv \Big|_a^b - \int_a^b v\, du$$

   List several integral forms which indicate this method should be used:

   $$\int x^n e^x dx, \int x^n \sin x\, dx, \int x^n \cos x\, dx, \int x^n \ln x\, dx, \int \ln x\, dx, \int e^x \sin x\, dx, \int e^x \cos x\, dx$$

   Work out an example below or on a new sheet of paper.

   $$\int xe^{2x}\, dx$$
   $$\begin{matrix} u = x & dv = e^{2x}\, dx \\ du = dx & v = \frac{1}{2} e^{2x} \end{matrix}$$

   $$\int xe^{2x}\, dx = x\left(\frac{1}{2} e^{2x}\right) - \int \frac{1}{2} e^{2x} dx = \frac{x}{2} e^{2x} - \frac{1}{4} e^{2x} + C = \frac{1}{4} e^{2x}(2x - 1) + C$$

Figure 5.10 Integration Strategy: A Second Swing at Products

Integration Strategy
A Second Swing at Fractions

When the integrand is a **fraction which cannot be simplified by the "First Swing" methods**:

1. If there is an **expression raised to a power**, try a ___u-substitution___ for the ___expression___.

    Work out an example:

    $$\int \frac{2}{(4-x)^{10}} dx \quad \begin{array}{l} u = 4 - x \\ du = -dx \end{array}$$

    $$= \int -2u^{-10} du = \frac{2}{9} u^{-9} + C = \frac{2}{9}(4-x)^{-9} + C$$

    $$= \frac{2}{9(4-x)^9} + C$$

2. If the ___numerator___ is the derivative of the ___denominator___ (or a multiple thereof), try a ___u-substitution___ for the ___denominator___.

    Work out an example:

    $$\int \frac{3x^4}{x^5 - 1} dx \quad \begin{array}{l} u = x^5 - 1 \\ du = 5x^4 dx \end{array}$$

    $$= \int \frac{3}{5} \frac{1}{u} du = \frac{3}{5} \ln|u| + C = \frac{3}{5} \ln|x^5 - 1| + C$$

3. If the integrand is an **improper rational function** (i.e. a fraction of ___polynomials___, with the degree of the numerator ___greater than___ the degree of the denominator), perform ___polynomial division___, then reassess the integrand.

    Work out an example:

    $$\int \frac{x^3 + 3x^2 + 2x + 3}{x + 1} dx = \int \left(x^2 + 2x + \frac{3}{x+1} \right) dx = \frac{1}{3}x^3 + x^2 + 3\ln|x + 1| + C$$

Figure 5.11 Integration Strategy: A Second Swing at Fractions

> Integration Strategy
> Exponential Functions
>
> When the integrand contains an **exponential function**:
>
> 1. Try a u-substitution for the ___exponent___.
>
>   Work out an example with the natural exponential:
>
> $$\int e^{3x+1} dx \quad \begin{array}{l} u = 3x+1 \\ du = 3dx \end{array}$$
>
> $$= \int \frac{1}{3} e^u du = \frac{1}{3} e^u + C = \frac{1}{3} e^{3x+1} + C$$
>
>   Work out an example with an exponential of base $a \neq e$:
>
> $$\int x 7^{2x^2-4} dx \quad \begin{array}{l} u = 2x^2 - 4 \\ du = 4x\, dx \end{array}$$
>
> $$= \int \frac{1}{4} 7^u du = \frac{1}{4\ln 7} 7^u + C = \frac{1}{4\ln 7} 7^{2x^2-4} + C$$
>
> 2. Try a u-substitution for the ___entire exponential___.
>
>   Work out an example with the natural exponential:
>
> $$\int \frac{e^x}{e^x + 1} dx \quad \begin{array}{l} u = e^x + 1 \\ du = e^x\, dx \end{array}$$
>
> $$= \int \frac{1}{u} du = \ln|u| + C = \ln|e^x + 1| + C = \ln(e^x + 1) + C$$
>
>   Work out an example with an exponential of base $a \neq e$:
>
> $$\int x^3 3^{x^4} \left(3^{x^4} + 2\right)^5 dx \quad \begin{array}{l} u = 3^{x^4} + 2 \\ du = 3^{x^4}(\ln 3)(4x^3)\, dx \end{array}$$
>
> $$= \int \frac{1}{4\ln 3} u^5 du = \frac{1}{24\ln 3} u^6 + C = \frac{1}{24\ln 3} \left(3^{x^4} + 2\right)^6 + C$$

Figure 5.12 Integration Strategy: Exponential Functions

encourages thought behind each process. While a few subtopics are added in these revised sheets, the bulk of the added length is due to the space allotted for students to fill in information and to provide and work out examples. As with the graphing techniques handouts, suggested responses are provided on the worksheets. Blank versions of these sheets, their keys, and a couple of bonus worksheets are available at www.routledge.com/9780367429027.

Integration Strategy
Logarithmic Functions

When the integrand contains a **logarithmic function**:

1. Try a u-substitution for <u>the entire logarithm</u>.

    Work out an example with the natural logarithm:

    $$\int \frac{\ln x}{x} dx = \int u\, du = \frac{1}{2}u^2 + C = \frac{1}{2}(\ln x)^2 + C$$

    $u = \ln x$
    $du = \frac{1}{x}dx$

    Work out an example with a logarithm of base $a \neq e$:

    $$\int \frac{\log_3 x}{x} dx = \int (\ln 3)\, u\, du = \frac{\ln 3}{2}u^2 + C = \frac{\ln 3}{2}(\log_3 x)^2 + C$$

    $u = \log_3 x$
    $du = \frac{1}{x \ln 3}dx$

2. Try a u-substitution for <u>the expression in the logarithm</u>.

    Work out an example with the natural logarithm:

    $$\int \ln(4x + 1)\, dx \quad \begin{array}{l} u = 4x + 1 \\ du = 4\, dx \end{array}$$

    $$= \int \frac{1}{4} \ln u\, du = \frac{1}{4}u(\ln u - 1) + C = \frac{1}{4}(4x + 1)(\ln(4x + 1) - 1) + C$$

    (Integration by parts is used to compute $\int \ln u\, du$; work omitted above.)

    Work out an example with a logarithm of base $a \neq e$:

    $$\int \log_2(7x - 2)\, dx \quad \begin{array}{l} u = 7x - 2 \\ du = 7dx \end{array}$$

    $$= \int \frac{1}{7} \log_2 u\, du = \frac{1}{7}u\left(\log_2 u - \frac{1}{\ln 2}\right) + C = \frac{1}{7}(7x - 2)\left(\log_2(7x - 2) - \frac{1}{\ln 2}\right) + C$$

    (Integration by parts is used to compute $\int \log_2 u\, du$; work omitted above.)

Figure 5.13 Integration Strategy: Logarithmic Functions

Integration Strategy
A First Swing at Trigonometric Functions

When the integrand contains a **trigonometric function**:

1. Try a *u*-substitution for ___the angle___.

    Work out an example:

    $\int \sin(6x)\, dx \quad \begin{array}{l} u = 6x \\ du = 6\, dx \end{array}$

    $= \int \frac{1}{6} \sin(u)\, du = -\frac{1}{6}\cos(u) + C = -\frac{1}{6}\cos(6x) + C$

2. If the integral contains **more than one trig function**:
    Try a *u*-substitution for ___an entire trig function___.

    Work out an example:

    $\int \sin(3x)\cos(3x)\, dx \quad \begin{array}{l} u = \sin(3x) \\ du = 3\cos(3x)\, dx \end{array}$

    $= \int \frac{1}{3} u\, du = \frac{1}{6} u^2 + C = \frac{1}{6}\sin^2(3x) + C$

    Or

    $\int \sin^2(3x)\cos(3x)\, dx \quad \begin{array}{l} u = \sin(3x) \\ du = 3\cos(3x)\, dx \end{array}$

    $= \int \frac{1}{3} u^2\, du = \frac{1}{9} u^3 + C = \frac{1}{9}\sin^3(3x) + C$

3. If the integral involves **secant, cosecant, tangent, and/or cotangent and the above methods have been unsuccessful**, try rewriting the integrand in terms of ___sine and/or cosine___.

    Work out an example:

    $\int \tan x\, dx = \int \frac{\sin x}{\cos x}\, dx \quad \begin{array}{l} u = \cos x \\ du = -\sin x\, dx \end{array}$

    $= \int -\frac{1}{u}\, du = -\ln|u| + C = -\ln|\cos x| + C$

    (Equivalently, this answer can be written as: $\ln|\sec x| + C$.)

Figure 5.14 Integration Strategy: A First Swing at Trigonometric Functions

Integration Strategy
A Second Swing at Trigonometric Functions

When the integrand contains a **power(s) of trigonometric function(s) and the integral cannot be solved using the "First Swing" methods**:

1. Integrals consisting of **sine and/or cosine terms**:
   - If **the power on either** sine or cosine **is odd**, pull off one copy of the trig function raised to the __odd__ power and convert the remaining copies to the other trig function, using the identity __$\sin^2 x + \cos^2 x = 1$__. If both powers are odd, pick __either one__ to convert, with preference to one raised to a __lower__ power.

     State two examples, then complete the problems on a separate sheet of paper:
     $\int \sin^3(8x) \cos^2(8x)\, dx$, $\int \cos^5 x\, dx$

   - If **both** are **raised to even powers**, convert __all terms__ to __cosine terms__ using one or both of the identities:

     $\sin^2 x = \frac{1}{2}(1 - \cos(2x))$ and $\cos^2 x = \frac{1}{2}(1 + \cos(2x))$

     State two examples, then complete the problems on a separate sheet of paper:
     $\int \sin^2(4x) \cos^2(4x)\, dx$, $\int \sin^6 x\, dx$

2. Integrals with **tangent and/or secant terms**:
   - If the power on secant is __even__, pull off one copy of __$\sec^2 x$__, and convert the rest of the secants to __tangent terms__, using the identity __$\sec^2 x = 1 + \tan^2 x$__.

     State an example, then complete the problem on a separate sheet of paper:
     $\int \sec^4 x\, dx$

   - If both tangent and secant are present and the power on tangent is __odd__, pull off one copy of __secant__ and one copy of __tangent__, and convert the rest of the __tangent terms__ to __secant terms__, using the identity __$\tan^2 x = \sec^2 x - 1$__.

     State an example, then complete the problem on a separate sheet of paper:
     $\int \sec^3 x \tan^5 x\, dx$

   - If neither is true, then __experiment__.

Figure 5.15 Integration Strategy: A Second Swing at Trigonometric Functions

3. Integrals with **cotangent and/or cosecant terms**:
   - If the power on cosecant is __even__, pull off one copy of __$\csc^2 x$__, and convert the rest of the __cosecant terms__ to __cotangent terms__, using the identity __$\csc^2 x = 1 + \cot^2 x$__.

     State an example, then complete the problem on a separate sheet of paper:
     $$\int \csc^4 x \cot^2 x \, dx$$

   - If both cotangent and cosecant are present and the power on cotangent is __odd__, pull off one copy of __cosecant__ and one copy of __cotangent__, and convert the rest of the __cotangent terms__ to __cosecant terms__, using the identity __$\cot^2 x = \csc^2 x - 1$__.

     State an example, then complete the problem on a separate sheet of paper:
     $$\int \csc^5 x \cot^3 x \, dx$$

   - If neither is true, then __experiment__.

Figure 5.15 Continued

Integration Strategy

A Second Swing at Powers: Square Roots of Quadratics

When the integrand contains **a square root of a quadratic and u-substitution fails**:

1. If the root is of the form $\sqrt{a^2 - x^2}$:
   - Use the substitution __$x = a \sin \theta$__ with the restriction: __$-\frac{\pi}{2} \leq \theta \leq \frac{\pi}{2}$__
   - Use the identity __$\cos^2 x = 1 - \sin^2 x$__ to simplify the root.

2. If the root is of the form $\sqrt{a^2 + x^2}$:
   - Use the substitution __$x = a \tan \theta$__ with the restriction: __$-\frac{\pi}{2} < \theta < \frac{\pi}{2}$__
   - Use the identity __$\sec^2 x = 1 + \tan^2 x$__ to simplify the root.

3. If the root is of the form $\sqrt{x^2 - a^2}$:
   - Use the substitution __$x = a \sec \theta$__ with the restriction: __$0 \leq \theta < \frac{\pi}{2}$__ or __$\pi \leq \theta < \frac{3\pi}{2}$__
   - Use the identity __$\tan^2 x = \sec^2 x - 1$__ to simplify the root.

Work out examples of each on additional paper.

Figure 5.16 Integration Strategy: A Second Swing at Powers: Square Roots of Quadratics

Integration Strategy
A Third Swing at Fractions

When the integrand is a **fraction which cannot be simplified by the "First Swing" or "Second Swing"**:

If the integrand is a **proper rational function** (i.e. a fraction of __polynomials__ , with the degree of the numerator __less than__ the degree of the denominator), try the method of __partial fractions__ then reassess the integrand.

Factor the denominator completely, then use the following guidelines. For each example requested, you need only state a partial fractions decomposition. You do not need to solve for the constants or integrate your example.

- If the denominator factors into distinct linear factors, the partial fractions will each have a __constant__ in the numerator. Provide an example of such an integrand and the form of its partial fractions decomposition:

$$\frac{x}{(x-1)(x+2)} = \frac{A}{x-1} + \frac{B}{x+2}$$

- If the denominator contains a repeated linear factor, the partial fractions will each have a __constant__ in the numerator and __each power up to__ the power of the repeated factor must be represented in the list of partial fractions. Provide an example of such an integrand and the form of its partial fractions decomposition:

$$\frac{x}{(x-1)(x+2)^3} = \frac{A}{x-1} + \frac{B}{x+2} + \frac{C}{(x+2)^2} + \frac{D}{(x+2)^3}$$

- If the denominator contains a distinct irreducible quadratic factor, the partial fraction in the decomposition will have a __linear expression__ in the numerator. Provide an example of such an integrand and the form of its partial fractions decomposition:

$$\frac{x}{(x^2+1)(x+2)} = \frac{Ax+B}{x^2+1} + \frac{C}{x+2}$$

- If the denominator contains repeated irreducible quadratic factors, the partial fractions in the decomposition will each have a __linear expression__ in the numerator and __each power up to__ the power of the repeated factor must be represented in the list of partial fractions. Provide an example of such an integrand and the form of its partial fractions decomposition:

$$\frac{x}{(x^2+1)^2(x+2)} = \frac{Ax+B}{x^2+1} + \frac{Cx+D}{(x^2+1)^2} + \frac{E}{x+2}$$

Notes:
a) When integrating partial fractions decompositions, we frequently need to use one of the following:
   - The __power__ rule
   - $\int \frac{1}{u} du =$ __$\ln|u| + C$__
   - $\int \frac{1}{x^2+a^2} dx =$ __$\frac{1}{a}\tan^{-1}\left(\frac{x}{a}\right) + C$__

b) These integrals will often have answers involving the natural log. Use log rules to __combine terms__ .

Figure 5.17 Integration Strategy: A Third Swing at Fractions

Sophomore Calculus

Since many students pursuing calculus past the first year may be either minoring or majoring in math or a math-related discipline, it becomes even more essential that students are responsible for understanding how and why techniques work. The content and order of material taught in sophomore calculus will vary based on the institution and text used for the course. This section will discuss a variety of opportunities you have with the topics that may occur in such a course.

Opportunities for Derivation

Deriving the formulas for arc length and surface area of a solid of revolution offer opportunities for different strategical approaches. Arc length can be done quite directly, while you might build conceptually to the derivation for surface area. Each can serve to improve the comprehension of the formula being derived as well as cement older concepts.

Deriving the formula for arc length utilizes previously learned theorems and procedures, allowing an opportunity for review and application. You will use the Pythagorean Theorem, with which students should be very comfortable, and you will revisit and apply the Mean Value Theorem. This proof also practices the concept of breaking down a region into manageable localized segments to determine what occurs globally for the curve, as done for area and volume.

In deriving the formula for the surface area for a solid of revolution, you might consider a gradual approach. For instance, you can start with an example of a right circular cylinder. Showing how this cylinder can be cut up into the rectangle of height h and length $2\pi r$, helps students comprehend the origin of the surface area formula. Then you can move on to a similar object, such as a parabola revolved around its axis of symmetry. Slicing this object into bands and cutting a sample band just as you did the cylinder, you arrive at the same rectangle as before of height h (the arc length of your curve) and length $2\pi r$. Completing this derivation by finding the formulas for h and r demystifies the formula and encourages your students to avoid blind memorization.

By participating in these derivations, students are given the opportunity to comprehend the origin of formulas and potentially retain them more easily. They may even begin to try to dissect new formulas they encounter or revisit those they have used for years. The more involved your students are in the process of deriving these formulas, whether by class discussion or a structured group exercise, the more effective the process will be in aiding their comprehension, insight, and recall.

Opportunities for Fixing What Never Looked Broken

Parametric equations may be your students' first exposure to any graphing that is outside the Cartesian box. They may find this tedious since they already know a method of graphing, so motivate the discussion with the desire to graph rambling, non-functions, such as those representing a flightpath of a bug. Another motivating question to consider is how one would represent traveling a path repeatedly (such as around a racetrack). Leading with such illustrations of how our tried-and-tested graphing techniques are lacking, creates the understanding that these new methods increase our capabilities. It also serves to pique the students' interest as to how one might achieve such tasks. This curiosity improves their ability to learn and recall the material.

Polar coordinates provide another opportunity to illustrate how restrictive our previous graphing abilities were. You can motivate your discussion by drawing a cardioid, three-leaved rose, or open washer and asking your students to think about how they would write an equation to represent the graph. Illustrating how polar coordinates can simplify expressions, even for just a circle, may also appeal to your students. They can sometimes be so overwhelmed with the new material that they miss the beauty of it, so consider taking avenues by which you can help them discover and appreciate this.

Opportunities for Building on (and Improving) Previous Knowledge

When you revisit content that students learned in a previous course, they often begin to understand it on a new level. During the original period of learning, students were taking in many separate pieces of the material, attempting to simultaneously process and assemble them into a cohesive picture. Bits of material that they knew, such as the how the sign of a derivative related to the graph of a function, can become more conceptualized and less of a memorized rule when revisited and expanded upon.

When you discuss partial derivatives, you can utilize the students' previous derivative knowledge in several ways. For instance, you can ask students to recall information about the tangent lines found in the first-year differential calculus and have them discuss parallels to the tangent planes you may now wish to produce. Similarly, you can have students recall the chain rule for one variable before discussing the rule for a function of two variables. A third opportunity arises with the second partials test. While some students may not have learned the second derivative test in their first-year calculus course, it is easy enough for students at this level to grasp. You might consider teaching it, if they have not seen it, because having that foundational material may make it easier for your students to see the concepts behind the second partials test. You could ask your students to conjecture how the second derivative might extend in the multivariable setting. Drawing each of these parallels serves to practice the retrieval of the old material and improve the comprehension of the new.

Opportunities for Opening and Closing Exercises and Handouts

When teaching sequences and series, one of the obstacles is simply getting your students to understand the difference between the two, oddly enough. After each has been defined in your course, consider an opening or closing exercise asking your students to distinguish between them. Can they further describe a relationship between the two? You could also ask students to define the terms monotonic and bounded and *why* these may tell us something about convergence or divergence of a sequence. Having students discuss these ideas outside of a specific problem they are trying to solve, helps isolate the concept at hand and avoid the attempt to memorize.

Just as with integration techniques, sometimes students will just blindly try to apply tests for series convergence instead of thinking critically about the process. Taking time to discuss why these tests work, versus simply stating the tests and applying them, is a first step. Consider approaching them as shown in the *Integration Strategy Handouts*, by constructing a similar fill-in-the-blank worksheets for convergence tests which makes students think about how to approach a series.

Elementary Linear Algebra

We expect to largely be dealing with students in math-related majors and minors in this course, so presumably they already enjoy the subject. In this course, you have the opportunity to help them realize how much more there is to this fascinating world of mathematics than the coordinate plane! Once again, using what they already know and understand can be a useful tool to foster a deeper understanding of the new concepts here. A distinct challenge with this course arises in presentation, such as when discussing extensive axiom lists and performing lengthy matrix calculations.

As tedious as it may be, I strongly encourage you to put all details of matrix operations and row reduction on the board for a while. It is not that the concept will necessarily be challenging for your students to follow. The difficulty arises when you and the class only *say* some steps and a few students miss it. They may quickly become lost and panicked or simply fall behind the discussion while discerning what was done. In addition, when the class looks back at examples later, students may not remember what transpired in a step.

An example demonstrating *full* details is provided Figure 5.18. You should be able to wean the details written in curly brackets as students practice the material and become comfortable performing these operations in their heads, but the directions written above the arrows are important information to maintain.

Requiring that students use similar notes as those written above the arrows when performing their own work will greatly aid your ability to follow their work and award partial credit. Combining steps in the first lesson may be too confusing, but after that consider asking your class if a particular combination is okay. They do not want to write out more than is necessary either and this question will help you feel out where they stand. While it is understandable to want to eliminate as much tedious note taking as possible, clarity should not be sacrificed for speed.

Opportunities for Handouts

Due to their length and complexity, it is challenging to provide shortcuts to ease the writing in the aforementioned matrix problems. One option for in-class matrix exercises is a handout with repeated rows such as the following, with ample space left between rows for student to complete calculations:

Several samples of such a handout are available at www.routledge.com/9780367429027. While this only eliminates a small amount of writing, it provides a framework for the in-class problems. Students may focus solely on the operations and computations at hand. Mirroring the layout of your handout on the board offers additional assistance for students in comparing

Find the inverse of $A = \begin{bmatrix} 1 & 8 & 0 \\ 0 & 4 & 2 \\ 0 & 0 & 2 \end{bmatrix}$.

$\left[\begin{array}{ccc|ccc} 1 & 8 & 0 & 1 & 0 & 0 \\ 0 & 4 & 2 & 0 & 1 & 0 \\ 0 & 0 & 2 & 0 & 0 & 1 \end{array}\right] \xrightarrow{-2R_2+R_1 \to R_1} \left[\begin{array}{ccc|ccc} 1 & 0 & -4 & 1 & -2 & 0 \\ 0 & 4 & 2 & 0 & 1 & 0 \\ 0 & 0 & 2 & 0 & 0 & 1 \end{array}\right]$

$\left\{\begin{array}{cccccc} -2R_2 & 0 & -8 & -4 & 0 & -2 & 0 \\ +R_1 & 1 & 8 & 0 & 1 & 0 & 0 \\ \hline \text{new } R_1 & 1 & 0 & -4 & 1 & -2 & 0 \end{array}\right.$

$\xrightarrow[-R_3+R_2 \to R_2]{2R_3+R_1 \to R_1} \left[\begin{array}{ccc|ccc} 1 & 0 & 0 & 1 & -2 & 2 \\ 0 & 4 & 0 & 0 & 1 & -1 \\ 0 & 0 & 2 & 0 & 0 & 1 \end{array}\right]$

$\left\{\begin{array}{cccccc} 2R_3 & 0 & 0 & 4 & 0 & 0 & 2 \\ +R_1 & 1 & 0 & -4 & 1 & -2 & 0 \\ \hline \text{new } R_1 & 1 & 0 & 0 & 1 & -2 & 2 \end{array}\right.$

$\left\{\begin{array}{cccccc} -R_3 & 0 & 0 & -2 & 0 & 0 & -1 \\ +R_2 & 0 & 4 & 2 & 0 & 1 & 0 \\ \hline \text{new } R_2 & 0 & 4 & 0 & 0 & 1 & -1 \end{array}\right.$

$\xrightarrow[\frac{1}{2}R_3 \to R_3]{\frac{1}{4}R_2 \to R_2} \left[\begin{array}{ccc|ccc} 1 & 0 & 0 & 1 & -2 & 2 \\ 0 & 1 & 0 & 0 & 1/4 & -1/4 \\ 0 & 0 & 1 & 0 & 0 & 1/2 \end{array}\right]$

$\left\{\begin{array}{l} \frac{1}{4}R_2 = 0 \quad 1 \quad 0 \quad 0 \quad 1/4 \quad -1/4 \quad = \text{new } R_2 \\ \frac{1}{2}R_3 = 0 \quad 0 \quad 1 \quad 0 \quad 0 \quad 1/2 \quad = \text{new } R_3 \end{array}\right.$

$A^{-1} = \begin{bmatrix} 1 & -2 & 2 \\ 0 & 1/4 & -1/4 \\ 0 & 0 & 1/2 \end{bmatrix}$

Figure 5.18 Sample work for finding the inverse of a matrix

work and copying notes. If you use the same format on quizzes and tests, it will add an air of familiarity and it may clarify the level of supporting work you expect.

In handling the axiom content in your course, you may find it beneficial to find an alternative to simply having students copy down long lists. If you are presenting a list of axioms on the board or on an overhead, you might refer your students to their text or handout to keep their focus on what the axioms *say* rather than on copying them. Just as with the graphing and integration review sheets, there is value in actually writing out axioms instead of simply reading them, so encourage students to write out important axioms, possibly on a fill-in-the blank worksheet. You could either select the most important axioms for students to write out in their entirety or have students fill in crucial points in each axiom.

Opportunities for Opening Exercises and Building on Previous Knowledge

While linear algebra opens a new door to the expansive world of mathematics, the foreign nature of the material can be intimidating. Using material students understand well to introduce a new concept can help demystify it. Contrasting new spaces to those students already know can deepen their understanding of the role axioms play.

When introducing the topic of computing matrix inverses, consider first discussing the multiplicative inverse of a real number. This is something they understand quite well, so they should be able to articulate their knowledge and defend their answers. Perhaps you could open class by asking them to state the multiplicative inverse of a specific real number and assert why their answer is correct. You could follow this by asking what condition must be met for x to be an inverse of y, where x and y are real numbers. And finally, do all real numbers have such an inverse?

Students already understand that the inverse of the real number 0 does not exist. When they approach the conversation on matrix inverses with this thought fresh in their minds, it seems perfectly reasonable that the inverse of a matrix might not always exist. Additionally, the problem of dividing by zero relates both to the question of an inverse in the set of real numbers and to the matrices with determinant zero. Drawing on this parallel with the real numbers makes the new concept of an inverse matrix more approachable.

When introducing the concept of a vector space, you can first analyze spaces with which they have worked for *years*, such as $\mathbb{R}$ and $\mathbb{R}^2$, with the usual addition and scalar multiplication. You can open class by asking students to discuss a property of one of these comfortable spaces, and then ask them to give an example of another space discussed in class which shares this property. Similarly, what is a space that does not have this property? For instance, you might ask students to verify that $\mathbb{R}$ is closed under multiplication and to provide an example of another space which meets this criterion. What is an example which is not closed under multiplication? Where multiplication might not even be defined? The parallels allow students to see the similar spaces as less intimidating, and the contrasts help to further understand the complexities of these spaces and the axioms we are attempting to satisfy.

Proof Courses

A proofs course may be the first exposure your students have to a non-computational view of mathematics. Students who have never needed much assistance in the past may require your guidance to understand the shift in strategy. Just as you can have a hierarchy in approaching integration techniques or tests for convergence of a series, you can help students approach a proof by organizing the process. There is an excellent opportunity to encourage well-crafted proofs by encouraging revision and analysis of methodology.

Opportunities for Handouts

Consider crafting fill-in-the-blank handouts for proof methodology which encourages students to use some preliminary analysis. Certainly, some problems call out for a particular method, such as contradiction or induction, and that should be considered before proceeding through the list of methods you are exploring. Just as with the integration handouts, these can gradually build up for each new technique. Once you have added enough techniques, you might create a sheet that helps students distinguish between their available choices. Utilizing cumulative quizzes may serve to keep all of the options for proof strategy fresh in your students' minds.

Opportunities for Opening and Closing Exercises

Consider opening class by posting a problem that students may not possess the knowledge to solve. For instance, the question might include a term or function you have not yet defined, so they cannot proceed with an actual proof just yet. Ask them to predict the appropriate strategy and explain their choice. For instance, the form of the question might suggest that a proof by contradiction or proof by induction is a likely avenue. By withholding some information, you require the class to think about the underlying approach rather than blindly stabbing at the proof.

If you are short on time, a quick but valuable exercise is to have students write out the meaning of quantifiers or symbols you have been using in class. While the mathematical shorthand symbols $\forall, \exists, \ni$, and $\therefore$ are excellent tools for saving time in both lecture and proof attempts, students may confuse when to use each quantifier and also confuse what the symbols represent. Consider asking them to write down the quantifier that represents, say "for all," as well as asking them to write out the words represented by a given symbol. You can provide a short proof which is chock full of symbolic notation and ask the class to "translate" it into words alone.

A longer activity that works well at the beginning or end of class is proof revision. You can present the class with a proof to correct or improve upon. Finding errors in the proof works on students' analytical skills and checks their comprehension of valid arguments. Comparing two sample proofs allows students to discover how omissions, misused quantifiers, or poor structure may weaken or falsify a proof. Cleaning up an accurate, but unnecessarily messy proof requires focusing on the critical components and encourages thoughtful proof writing. An example of a proof you could ask students to improve upon is one in which proof by contradiction was used unnecessarily. For instance, you might provide them with the following proof that the product of rational numbers is rational.

Let x and y be rational numbers. Then there exists integers m, n, p, and q, with n and q nonzero, such that $x = \frac{m}{n}$ and $y = \frac{p}{q}$. Suppose that xy is irrational. Then xy must not be a fraction of integers of nonzero denominator. But $xy = \frac{m}{n} \cdot \frac{p}{q} = \frac{mp}{nq}$ and since m, n, p, q are integers, mp and nq are integers. In addition, since neither n nor q is zero, nq must be nonzero. Thus, xy is a fraction of integers with nonzero denominator and we reach a contradiction.

You can ask students to identify which method was employed and to construct a proof using a different strategy. When students revise the above proof by contradiction, they can arrive at the much more appealing direct proof:

> Let x and y be rational numbers. Then there exists integers m, n, p, and q, with n and q nonzero, such that $x = \frac{m}{n}$ and $y = \frac{p}{q}$. Thus, $xy = \frac{m}{n} \cdot \frac{p}{q} = \frac{mp}{nq}$ and since m, n, p, q are integers, mp and nq are integers. In addition, since neither n nor q is zero, nq must be nonzero. Hence, xy is rational.

The direct proof is cleaner, shorter, and overall more pleasant to read, so students may begin to see the value of this revision in their own work. In general, if a direct proof can be accomplished, it generally is more illustrative of the concepts at hand than a proof by contradiction and this exercise helps to drive home that point.

Setting a Clear Standard

The process of writing proofs is probably quite new to your students. Providing a clear standard for how a graded proof should be written is important. Requiring that they omit symbolic quantifiers in graded responses, may help clarify precisely what students mean to say. In the same vein, requiring students to use complete sentences may assist in avoiding misunderstandings.

When students write a skimpy proof, it can be difficult to discern if they truly saw the crux of the proof or if they took leaps they did not understand to reach the desired conclusion. They may be of the mindset that their professor understands why a statement was important or why a conclusion was valid, leading them to omit necessary arguments. It may help to remind them that, most frequently, the result of a proof is already provided, and the task is to clearly demonstrate the path to reach that end. One option is to tell students that a fellow college student, who is not studying math, should be able to read and reasonably understand what they have written, given necessary definitions of terms.

You will likely want to use quantifiers in examples you write on the board. This eases the writing for you and your students and offers an opportunity to discuss the correct usage. If your write-up of a proof falls short of what you would accept for full credit on a graded exercise, you should clarify this for your class. Consider a closing exercise which would identify what modifications would need to be done to a proof presented in class, if it were to be submitted on a graded paper.

Upper-Level Courses

These courses are often fun and refreshing to teach. The students' interests in math are sufficient to push them into pursuing topics they do not understand, and the level and uniqueness of the material makes the lessons more intriguing. These courses can be quite captivating for your majors, but intimidating for them as well.

One of the especially enjoyable aspects of a course of this nature is that you are showing students material that they have likely seen virtually none of before. That freshness is exciting after so many calculus classes! Not only are students listening with a different level of care but you most likely have a blank slate to work with in regard to notation, terminology, and methodology.

In these upper-level courses, you are potentially working with the best mathematics students in your program, but this course may be one of the first steps they have taken into highly theoretical material. In this case, start slow, especially if you do not have a firm sense of the students' abilities and background. It is much more desirable to increase the difficulty level if students are responding well to the material than to need to backtrack and recover from having overestimated their readiness. If you begin at a level that is too sophisticated, students may comprehend little of the foundations of the course and quickly become frustrated.

I once asked students to prove that the identity mapping was continuous on a topology quiz. My goal was to build confidence by giving them the opportunity to demonstrate their knowledge of proving continuity with a trivial map with which to work. Unfortunately, some students performed quite poorly. It was a reminder to me how foreign the concepts were to my class. The students who struggled with this problem were stunned to see the simplicity of the solution when we discussed it in the next class. The experience was eye-opening for them as well. The material, while quite different from anything they had seen prior, was in fact do-able! There was a notable shift in the students' approach and the future quizzes, though increasingly more challenging, were much improved.

Opportunities within Complex Variables

Building on Previous Knowledge

A course on complex variables offers ample occasions for drawing parallels and revisiting previously learning material. First-year calculus can be directly drawn upon when you approach limits, the limit definition of the derivative, and differentiation formulas. Sophomore-level calculus courses provide the background in sequences, series, convergence/divergence, and polar coordinates. Vector calculus or linear algebra introduces students to vectors, which is helpful as you begin graphing.

The only snag in drawing on the earlier courses is that some students may have forgotten the material learned one to three years ago. In each of the above examples, consider revisiting the original content before discussion begins for complex variables. This allows you to reinforce the earlier knowledge and make the current material more accessible. If you close class by asking students to compare and contrast the content in the setting of the real numbers versus complex you can solidify their earlier understanding and create a deeper comprehension of the new.

Something New, Something Old (but Improved!)

A fun feature of a course on complex variables is the ability to do something different and yet not too advanced or difficult. For example, graphing with exponential form, integration along contours, and roots of complex numbers are all concepts that can be challenging at first, but appealing in their uniqueness. They each present an opportunity to show your enthusiasm for the beauty of mathematics.

Revealing simple approaches to problems that were quite challenging in calculus gives a practicality to this material. For example, applying the residue theorem to an integral for which we cannot find the antiderivative helps to expose students to a practical value of this subject. Motivating the discussion with an opening exercise which asks students to compute such an integral may help students appreciate the elegance of the new material.

Growth through Evaluation and Education   6

When I sat down to write my teaching statement for my tenure dossier, I started by opening one written for my third-year review. The same statement had been revised a number of times over the years for job applications and my first- and third-year reviews. I was somewhat bored as my eyes moved over the first paragraph for yet another revision until they stumbled upon statements with which I no longer entirely agreed. I saw that further analysis and explanation was needed to accurately describe the nuances of my philosophy.

As I began writing, I realized my perspective had modified over the years in ways so subtle that I had not taken notice. I sometimes struggled to recall when I had changed certain strategies and what had motivated these changes. In this section, I outline a formal method by which you can not only build up a healthy supply of feedback but also document the process to provide a record of exactly how and why you made alterations in your classroom. The more consistent and detailed you make your evaluation process, the easier and more useful documentation of your growth becomes.

While I will presume moving forward that the general reader is pursuing a career in academia, readers who only plan to teach for the immediate future can benefit by utilizing the tools presented here. The more successful you make your classroom experience for your students, the more enjoyable the time will be for *you* as well. Applying techniques to record problematic issues in your classroom and identify changes you want to make need not take considerable time.

Throughout your career you will hopefully make many adjustments, try new approaches, and address concerns expressed to you by superiors, colleagues, and students. As you prepare to apply for jobs, encounter reviews, and apply for tenure or promotion, it will be increasingly important to have a definitive approach in the classroom. The process of persistently analyzing your teaching skills and striving to improve your techniques will not only serve to enhance your abilities in the classroom but also to develop your capacity to communicate your philosophies to others.

To be able to fully illustrate your growth, it will be important to record your challenges and improvements. Creating the practice of documenting your teaching and its development will help you become a better, more focused, and efficient professor, and aid in creating dossiers for various applications. This process can be initiated quite simply in your day-to-day experiences of teaching and will naturally evolve into a continual refinement of your craft. This chapter provides fill-in sheets by which you can reflect on your strategies and request evaluations from students and peers in an organized and consistent manner. Copies of these sheets are available at www.routledge.com/9780367429027.

If you are concerned about revealing failures in a teaching dossier, don't be. My husband was pleasantly surprised when a member of his tenure-review committee mentioned to him after his review was complete that he appreciated his honesty and frankness when addressing the challenges he had faced and that he had made no attempt to ignore, hide or paint them rosy. Instead, he acknowledged each and spoke to how he had worked to address issues.

Every teacher has faced challenges. How you meet and address them determines your quality as a teacher. During a tenure review, your committee will have access to the formal reviews and student evaluations. Rather than deny the validity of all concerns raised, acknowledge the areas which needed improvement and the steps you took. Consider the advice from Joy Burnham, Lisa Hooper, and Vivian Wright in their *Faculty Focus* article on creating annual tenure and promotion dossiers:

> Hiding problem areas in your dossier instead of noting them will not assist you and, instead, can be detrimental to your success. Similarly, embellishing aspects of your dossier is almost always viewed unfavorably. Transparency, and honesty, in representing all aspects of your teaching … is important in all cases.
>
> (2012)

As you proceed with the process of evaluating and documenting your growth, try to do so with openness to critique and suggestions. You may not agree with everything that you hear or read but make every effort to apply objective thought to criticisms and concerns. You will gain a stronger sense of your teaching philosophy when you recognize deficiencies and implement suggestions, as well as when you are successfully able to justify your own course of action.

It is not my intent to outline the creation of your first teaching dossier, since this is years away for the anticipated audience of this book, but the practices I suggest will hopefully assist you in being prepared to assemble and produce such a collection. If you are currently teaching at an institution where you hope to pursue tenure, then touch base with your chair and the other administrators who will evaluate you at that time, regarding the materials that you need to collect. There should be clearly stated guidelines available as to the required elements but follow up with colleagues to learn what additional materials may be helpful. Reach out to recently tenured faculty to ask for advice on what data you may want for your dossier preparation.

The Stearns Center for Teaching and Learning at George Mason University hosts a detailed page, "Documenting Your Teaching," which offers guidance for creating a teaching portfolio. Among the recommended components for an effective teaching portfolio are the following (2017):

- evidence of a continuing effort to improve your teaching
- supporting evidence for your teaching statement, such as student and peer evaluations, unsolicited feedback from students or scholarly work
- explanations as to how these illustrate your described teaching philosophy

This chapter aims to provide you with these components as it lays out a variety of methods by which you can evaluate and improve your teaching and document your journey as you hone your practice and philosophy.

Self-Evaluation

While I have consistently participated in a good degree of self-evaluation, there are some more formal steps which, had I taken, would have made creating my tenure dossier less daunting. An informal method which I have always practiced is making daily notes on my lectures. This is not only the easiest place to start your self-evaluation, but an action which can have an immediate impact on your teaching.

In addition to day-to-day reviews, completing semester-end analysis of your teaching will help you sharpen your strategies and course content. You will begin to gain a sense of what you want to achieve in the classroom and how you best accomplish those goals. This will be helpful as you approach each new teaching endeavor and as you attempt to explain your philosophies to others. I have provided sample sheets which may help facilitate the self-evaluation process.

Post-Class Reflection

In *After Class* (Chapter 3), I suggested taking a moment after each class meeting to note what you felt worked well and what did not. This practice requires that you watch your students carefully as you lecture and facilitate discussion, tuning into what material they seem to grasp and what was confusing. Consider how questions arose in lecture and where students had difficulty comprehending your points or examples. Were they able to complete the in-class exercises you created and answer questions you posed?

Consider setting a time each week to make and review lecture evaluation notes. Ideally you would make at least some of these post-lecture notes immediately after class while thoughts are still fresh in your mind, but answering students' questions after class or rushing to another class or meeting may interfere with a timely review. Even when you have sufficient time to make immediate notes, there is some advantage to a secondary, weekly review.

After you have met with students in office hours and fielded questions in class, you may have a different perspective on how effective your chosen examples were. A weekly review allots time to not only catch up on days you missed but also address points you had not considered immediately after class. Similarly, if a subsequent formal assessment of student comprehension, such as a quiz, reveals gaps, you may have additional insight as to what concepts need attention. Perhaps you will determine that you need more varied examples or in-class problems or you may feel a different strategy might work better.

This process of noting problem areas need not take an extensive amount of time and it is not necessary to revise the lesson immediately. A few minutes to mark any errors or examples that were less illustrative than desired are enough. Even if you do not have immediate insight as to how you could rectify an issue, you may find you have a better perspective and fresh ideas the next time you teach the course. Once you have been through a course from start to finish a couple of times, you will have a more refined sense of what learning objectives you feel are most vital and potentially how you could best achieve them. Since you may need time to develop good solutions to issues that arise, it is helpful to write notes which are sufficiently clear and descriptive, so they may be deciphered at a later date.

While a quick note on the top of the day's lecture might suit your needs, you may find it useful to jot comments into a standardized sheet. This better enables you to see patterns of what is working well for this class and what is not. It also consolidates your notes into one spreadsheet or packet. You may find it easier to be consistent when you have a regular sheet to fill in, especially if these are kept handy to fill in between classes or when you return to your office. The sample sheet provided also provides a place to record the issues you realize later, after assessments or discussions with students have taken place. As you make post-class notes, keep the following points in mind.

Quick Glance: Making Effective Post-Class Notes

- *Keep it simple.* State the successes and challenges concisely. Try to get to the heart of the issue, if you can. If you are unsure of how to address a problem, reach out to colleagues for their perspectives.
- *Give yourself time.* Don't feel pressured to have an immediate solution. You may have better perspective on handling material in the future. Reach out to colleagues if you are unsure how to approach upcoming material.
- *Be clear.* Write your notes in such a way that you can return to them in several months and still understand the issue at hand.
- *Be consistent.* Try to make your initial notes as soon after class as possible. Set a weekly time to fill in any notes you may have missed.
- *Update.* After discussions with students and assessments, return to your notes to remark on where issues arose or to note the effectiveness of a new strategy.

Post-Class Reflection Worksheet

Post-Class Reflection Worksheet for _____			
Section number or topic	Explanations, examples, or activities which seemed most effective	Explanations, examples, or activities which need revision or correction	Notes from student questions, misconceptions, or errors

Post-Course Reflection

At the end of the semester, perform an overall review of the course. Ultimately, you will want to identify the top one or two major issues you would like to address in your next course. This is not to say you cannot make other small changes, but it is ineffective and unrealistic to attempt to work on everything you might improve upon at once. Making select alterations allows you to focus sharply on addressing concerns and more accurately assess whether the changes improved learning.

You can start a course overview by reading over your original list of learning objectives. Has it changed, solidified, or become more precise? Consider which topics or lessons went well and which need additional attention or revision. If you have control over the course content, are there any components you would choose to add or delete? Were there areas for which the lesson could be more interactive? Note if there were any unforeseen consequences of choices in notation or coverage that you prefer to avoid in the future.

Reflect on the pacing of the course and decide if you should make alterations in coverage or how class time is spent. Is there content you could explore in more depth or dwell on less in the future? Were there lessons for which the in-class activities took too long? Perhaps a revised structure to activities is warranted to improve pace or more activities are needed to help slow down a course that moved too quickly.

It is good practice to keep a sample of all handouts, overheads, quizzes, and exams for each course you teach. You may reuse some and use others as a basis for an improved version. Reviewing these papers at the end of the semester may reveal a gap you would like to fill the next time you teach the course. Note whether you want to revise the format or content of examinations or other forms of assessment. Even if you did not have a role in designing the assessments, you can analyze how well these are working for the course. You may determine you should add informal activities to better prepare your students for the formal assessments being implemented. It may be helpful to note your sense of the general skill level of the students enrolled in the particular section you are evaluating, as you might make different choices if the next class is more or less advanced.

If this is not the first time you taught the course and you made changes in a method of assessment or in classroom strategy, include an analysis of the perceived impact and any data you have collected to support this. It can be challenging to make direct comparisons when changes are significant, but consider how you can measure outcomes and what types of improvements would seem sufficient to support the change. There are undoubtedly some less quantifiable elements, such as student engagement, which may improve while not necessarily producing a significant difference in performance on major assessments. These enhancements are still relevant as both professors and students alike would prefer to be in a classroom which feels alive than one which is mundane.

If student evaluations directly impacted changes you made to a specific course, include a note to that effect in this file as well. What did students identify as an issue? How did you respond? Is there a change in student comments this semester to evidence success? Are there any new suggestions that have arisen? What is your reaction?

Another analysis to consider is how well you managed the organization of your course and its materials. Are there any changes that would assist you in being better prepared or more organized while in class? Is there anything you would do differently prior to the start of the next semester to organize assignments, handouts, or lesson plans?

As with the post-class reflection, you may find it useful to standardize your review process. This allows you to cross-reference between courses, note patterns, and better compare subsequent sections of the same course. The sample form provided could be expanded or simplified to suit your needs.

Quick Glance: Making Effective Post-Course Notes

- *Look at the big picture.* Overall, were the learning objectives and outcomes set appropriately? Were the learning objectives achieved?
- *Identify primary issues.* Do any learning objectives or outcomes stand out as needing revision or additional attention?
- *Examine strategies.* Were your in-class activities and opportunities for feedback effective? Were students properly prepared for the assessments employed?
- *Evaluate assessments.* Do you feel your methods of assessment accurately reflected the students' knowledge? Do you need more frequent or improved informal assessments?
- *Evaluate pace.* Did you often find yourself with extra time at the end of class or rushing to finish up? Were you able to maintain the intended course schedule? Where could you have added activities or altered the in-class approach to better pace the course? Was there content which should be added or removed?

- *Evaluate Changes.* Is there evidence that any changes you made resulted in improved learning or an improved student experience?
- *Update.* After reading student evaluations, return to your notes to reflect the aspects which reportedly helped or hindered their learning. Note any recommendations and whether they seem reasonable to employ.
- *Stay focused.* Identify one or two goals to work towards.
- *You don't have to know the answers!* Remember that the point of this process is to learn! Reach out to colleagues and mentors and discuss the issues you are facing. It often takes time to figure out how to make your approach more effective.

Post-Course Reflection Worksheet

Course _____ Term _____	
Most Effective Strategies	
Least Effective Strategies	
Topics which need a new approach, new examples, new exercises, new assessments, or: _____	
Should any learning objectives or desired learning outcomes be altered? If so, why?	
Which assessments should be altered and why? Were students prepared for the assessment style? Is there a need for improved or more frequent feedback?	

How was the pace for the course? Are new activities needed to slow the pace? Could activities be revised or done for homework to increase the pace?	
What are the three most common positive statements in the student evaluations? Do these reflect a change in strategy?	
What are the three most common concerns in the student evaluations? How should each be addressed? If no change is warranted, why not?	
If new strategies were implemented this term, were they effective? How was this evidenced?	
Were materials suitably organized? Were lessons well-prepared? Is any substantial new prep needed prior to the next iteration of this course?	
Overall assessment What aspects of your teaching do you want to focus on in the next course? (Choose one or two specific goals.) How will you measure the effectiveness of the proposed changes?	

Documentation of Your Self-Evaluation

Documentation of all the work described above provides you with a record of your challenges, your endeavors to improve, and your growth. To get the most out of all the reflection you have completed, consider creating a file for each course you teach which includes a copy of the comprehensive review you wrote for that course. Try to add to the file each time you teach the course. In doing so, you will have a concise record of the hurdles you faced and how you overcame them.

A thorough documentation process deepens your analysis, though you will likely not use every noted insight gained or modification made in crafting a teaching philosophy statement. You will probably identify key changes which stand out as most impactful, but you will be grateful for an ample written history of your progression as a teacher over the preceding years. The scrutiny which you applied will have sharpened your senses in the classroom and heightened your awareness of the student experience.

Peer Evaluations and Collaborations

In Chapter 3, I discussed how challenges in our lives outside the classroom, such as sleepless nights, can impact our presence in the classroom. It can therefore affect our performance evaluations as well as the quality of our self-assessment. Within a month of learning I had been tenured and right in the heart of a semester, my two-year-old daughter was diagnosed with autism. I can only assume that my cheerfulness in class was nonexistent, though the student evaluations did not report any concerns or complaints to this effect. Possibly, the classroom experience was less affected than I might have imagined because a class meeting was a small portion of my day when I could set my grief and uncertainty aside and do something which was innocuous and second-nature. My lack of clarity is exactly my point. There will be times when we do not accurately assess our own classroom environment or our performance in it, even when all is well in our personal lives. Utilizing and embracing the feedback from colleagues and students can be invaluable.

Selecting Diverse Observers

Research has shown that on-going formative feedback from colleagues and students is the best avenue to improved teaching (National Research Council, 2003: 76). Invite peers to visit your classes in addition to the observations that may be done by faculty in a supervisory role. It can be nerve-wracking to have a colleague observe your class and I cannot say that I have ever enjoyed such a visit, but the conversations afterward have always been positive. Visitors will hopefully highlight points they enjoyed about your approach as well as offer suggestions for alternatives. This type of observation can provide valuable information on content coverage and pedagogy, which you may not learn through student evaluations (Felder & Brent, 2016: 101).

I encourage you to seek ample feedback from math colleagues. Imperative in your feedback is how well you are achieving the learning objectives and demonstrating a mastery of the material. Feedback of this nature requires an observer well-versed in the content presented. A mathematics colleague may approach material from a different perspective or have another classroom strategy. Discussing your differences may spark new ideas or provide support for your current approach.

In addition to soliciting advice from faculty within your department consider inviting at least one faculty member outside the science, technology, engineering and mathematics (STEM) environment. Many institutions have formal options for a mentor or for peer review of teaching which you may find beneficial. Your overall demeanor, connection with students, and presence in the classroom are important components of your success. A professor who has minimal knowledge of the concepts you are covering will be potentially more tuned-in to these other aspects of your teaching. It is likely that your tenure-review committee will include a number of evaluators outside the STEM spectrum and gaining insight as to how such faculty might perceive you could prove valuable.

Collaborating with Your Colleagues

Discuss your approaches, successes, and failures with those who are teaching the same courses or have taught them in the past and seek out feedback on both the concrete and abstract elements of your teaching. When you are approaching a new course, ask those who have taught it before if there are any bits of advice on coverage they have to offer. If you are encountering a particular challenge in one of your classes, discuss with your peers whether they have faced a similar struggle and what tactics they attempted. After you have constructed drafts of your own, look at the handouts, in-class exercises, quizzes, and tests of your colleagues to see how their approaches may differ.

Documentation of Your Peer Evaluations and Collaborations

Proactively seeking out voluntary observations of your classroom demonstrates a dedication to improving your skills and providing the best experience for your students. Create a file to contain the results of any reviews or feedback you have received, as well as your goals for improvement. Such a record may provide some excellent examples in a teaching portfolio to mark your growth and your persistent reflection.

In your file, you can include formal review letters in which each reviewer will likely have documented the classroom discussion and your strengths and weaknesses as an instructor. If you meet with the reviewer after the observation, you could add to the file notes on any additional points mentioned in that discussion. Colleagues who casually observe a class to offer feedback may simply give you a few comments after class and not write up a review. You could request a written review or create your own, by making a record of all the reviewer's perceptions. Alternatively, you could provide the casual reviewer with a worksheet to fill in during or shortly after the class, such as the sample worksheet provided in the next section. While you could offer this to a formal reviewer, they will most likely include all of this information in a letter for you.

For any observation, consider writing up your own assessment of the class visited. You may not remember much about it later, so assess the general skill level of the students, their responsiveness, engagement, etc. Make notes regarding the day observed. How representative or typical was the class meeting? Note what you feel you did well, and what needed improvement. Were you implementing any new strategies? Record your immediate thoughts on the reviewer's observations and suggestions.

Occasionally, you may be lucky enough to have a truly remarkable conversation with a peer or superior in which you become enlightened or excited about how to approach a particular issue. If so, you might jot down some highlights to include in your file as well. As you review these documents in the future, you may find that such a conversation impacted you in a way that is relevant for a teaching statement.

Worksheet for Informal Peer Review

Course_____ Reviewer_____ Date_____	
What positive aspects of my teaching did you observe today?	

Where could you see a need for improvement?	
Could you see and read my board work?	
Could you hear me clearly?	
Did I engage students in the classroom discussion well?	
Did I answer questions effectively?	
Was the lesson effective overall?	

Please write any additional comments on the back. Thank you for your time today!

Student Evaluations

A few years ago, during a trip to the beach, my husband took our kids down to the water for an early morning swim. I was sitting out on the patio of our hotel room and called out to them as they were walking back across the lawn to our room, "so how was it?" Simultaneously, my son and husband replied, "pretty good!" and "that was *horrendous!*" My son had enjoyed care-free play, while my husband had desperately struggled to keep our curious daughter away from jellyfish and anxiously counted the minutes until he could end the excursion. They had both been at the same beach, mere feet away, but had completely opposite experiences. As we digest evaluations from our students, it is important to recognize that many factors affect the experience of those in our classroom. You are likely to see conflicting statements in course evaluations from two students sitting "mere feet away" in the same classroom, but it does not necessarily mean that either assessment is incorrect.

Evaluating Early

One of the most effective tools for improving your teaching may be feedback from informal evaluations gathered throughout a course and these permit mid-course corrections (National Research Council, 2003: 76). You may find it helpful to do mid-term evaluations in addition to the formal evaluations your institution will most likely perform at the end of the semester. Such surveys are useful even for an established professor, especially when teaching a new course or utilizing a new teaching strategy or course design. You can ask a variety of questions that are specific to your course and the tools you have chosen to use in the classroom. The mid-term evaluations give you the opportunity to address concerns or struggles that your students have before the course ends (Fink, 2003: 144). While this could be a single check-in at the midterm, there are a variety of options to receive more regular feedback with little to no class time consumed.

My first semester teaching as a faculty member, I included a section at the end of each weekly quiz where students would write concerns or requests for the course. While I initially feared that the stress of a quiz might elicit less-than-constructive comments, I am pleased to report that students used them effectively. Remarks gradually faded as we neared the midterm and the class became more comfortable discussing issues with me directly.

To remove the potential stress of assessment, you could utilize the minute paper. At the end of class, you can ask students to specifically speak to points of confusion (or clarity!) from the day's material (National Research Council, 2003: 79). Similarly, you can poll the class several times during the term as to what they feel is going well, what needs improvement and what suggestions they might have (National Research Council, 2003: 81). These polls need not take place in class, if time is especially tight. Instead, you could request students submit answers through an online class page or e-mail. You might also give students the option of submitting anonymous feedback. A sample mid-term evaluation follows, but a variety of more in-depth options for student evaluations can be found in the appendix of the National Research Council's *Evaluating and Improving Undergraduate Teaching* (a free pdf of which can be downloaded at the National Academies Press website, www.nap.edu).

Sample Mid-term Evaluation

Which aspects of this course do you feel best help you understand the material?	
Which aspects do you feel may interfere with learning?	
Do you have suggestions which could improve your experience in the classroom and/or your comprehension of the material?	
Do you feel the quizzes (*or substitute assignments, tests, etc.*) have accurately reflected your knowledge? If not, do you have a suggested adjustment or addition?	
Do you feel you have received useful feedback as to which errors you have made? Do you see how to improve your performance? If not, what would better assist you?	
Do you have any concerns not addressed above?	

Keeping an Open Mind

It can be both incredibly gratifying and fundamentally disappointing to read student evaluations. Some pedagogical choices may be unpopular because they are challenging. Try to avoid any temptation to dismiss complaints with the notion that students lack perspective on our goals as instructors. You may gain insight on how to improve their experience or simply realize the need to better explain the purpose of the challenges you put forth. While peers may offer more insightful feedback on classroom strategy, they may not accurately assess the student experience. Student evaluations give feedback on your approachability and how well you captured interest in a way that a colleague cannot (Felder & Brent, 2016: 101). Only students can express how invested you seemed in the class' learning, helping individuals, and understanding their questions. In *What the Best College Teachers Do*, Bain asserts:

> Part of being a good teacher (not all) is knowing that you always have something new to learn – not so much about teaching techniques but about these particular students at this particular time and their particular sets of aspirations, confusions, misconceptions, and ignorance. ... We will not reach all students equally, but there is something to learn about each one of them and about human learning in general.
>
> (2004: 174)

Your evaluations potentially present the last opportunity to learn about your students. How are students describing the learning environment and do they suggest ways in which your actions helped or harmed their learning? Identify the concerns students voiced that you could address in the future. Are there elements that are course-specific or do the concerns generalize?

After exhausting your efforts to objectively harness useful information, you may be left with comments you feel are unmerited. Student evaluations may contain bias and certain elements beyond your control could be affecting your students' review of your course. Required classes receive lower ratings than electives and math and science courses receive lower ratings than humanities and social science courses (Cashin, 1995b). Try to assume the best intentions of your students and be open to critique but realize that you may not be able to please everyone.

Documentation of Your Student Evaluations

When I began the process of writing my third-year review dossier, I re-read every student evaluation I had received up to that time. The thought of repeating this effort when it came time for my tenure review was daunting, so I began the practice of highlighting the key statements in the course evaluations as I read them at the end of each semester. When it came time to address evaluations in my tenure review, this practice allowed me to focus on the most descriptive, well-written comments and gloss over repeats of similar statements that were less quote-worthy. I also avoided re-reading isolated comments from a disgruntled student here or there. Note that I do not mean that in this process I only marked comments that sang my praises and would be a delight to read later. I highlighted concerns of a class that were voiced by more than one student as well as the occasional isolated concern that I thought was something I wanted to address in the future. At the time, I did not create any written analysis of each report, which in retrospect would have been useful.

Consider keeping a *select* record of students' comments in a file for student evaluations and your reflections on them. As you read through evaluations at the end of each semester, you can create a list of prevalent statements that reveal your achievements as well as those revealing areas of concern. You may want to describe the student population in the class, especially if it was atypical in any way that may have impacted the evaluations or how you conducted class. What are your immediate thoughts on the content of the evaluations? Describe any issues you want to address. If students request a change with which you disagree, explain the rationale for maintaining your current practice.

Creating a "highlights" sheet for each set of evaluations in which you summarize the overall feedback and your thoughts, as well as a list of direct quotes, will provide you with a concise snapshot of the student responses and any changes you may want to make. You may also want to include a record of any useful comments from students that arose during office-hour conversations or e-mails. A sample fill-in worksheet is provided which could be used as your record or to organize your thoughts towards a more detailed review.

Post-Course Student Evaluation Worksheet

Course _____ **Term** _____	
Reported Positive Aspects of Course	
Reported Areas of Concern	
Were there anomalies in the student population which may have affected responses?	
Were there any unusual aspects – time of day that class was offered, cancellations, student absences, etc.?	
What are the top three areas in which learning and/or the student experience could be improved?	
What actions can you take towards the goals listed above? How can you measure success?	

Pedagogical Professional Development

In the introduction to this book, I noted how I received essentially no training prior to teaching my first class. What few books on education I encountered did not translate well to the math classroom – or perhaps I lacked the insight to see how I might use the advice offered. I regret that I assumed such valuable advice simply did not exist for me and I did not continue to seek out educational resources until much later in my career. I encourage you to pursue reading in various formats (books, journals, blogs) and investigate the opportunities to attend seminars and workshops at your institution and online. Even if you do not have a teaching resource center on your campus, you will find many institutions post videos and webinars from their events online.

Documentation of Your Education

Creating an education file will allow you to keep track of any reading you do and events you attend. You can include the programs from events (or an e-mail which advertised it) and write a paragraph or two describing the topic and your reaction. Did you find the content interesting, exciting, or thought provoking? Will you plan to pursue the ideas presented? Note if you disagreed with the speaker or found the contents of the talk difficult to implement in your classes. If you read books, articles in journals, or blogs, include an entry in your education file for each with a synopsis of your reaction and interest in the content of each. Any concrete actions you can cite which document your initiative to learn more about your craft further evidences your commitment and dedication. Formulating and articulating your reaction to readings and events develops and enhances your personal philosophy.

Going beyond Traditional Lecture 7

Any instructor, regardless of experience, can benefit from learning about a variety of methods and strategies for the classroom. As you begin your teaching career, you will most likely gravitate towards modeling the type of classroom instruction you have experienced most. It is helpful to develop a sense of your student body and the courses you teach before attempting a strategy that is drastically outside your comfort zone. If you experiment with tools used in these approaches as you develop your ideas about achieving the learning objectives for your courses, you will be better prepared should you decide to design a course to more fully implement these strategies. This chapter will discuss some options and alert you to issues you should consider in advance of such a course design.

Preparing to Try Something New

I am *not* a runner but let us suppose after watching a marathon I (somehow) became inspired to attempt one. It would hardly be advisable to just sign up for a 26.219-mile run without training first. Neither would it be sufficient to simply *read* about the recommended techniques for long-distance running. While researching the best ways to train and talking to experienced marathoners would reap benefits, at some point, I would have to start with a jog around the block.

If you are considering a new course design, try out activities that would be compatible. In *After Class* (Chapter 3) and *Post-Class Reflection* (Chapter 6), I discussed making notes about what worked well in a class and what did not. You can use this same approach of daily note-making to keep a look out for opportunities as you conduct class. Building up a list of active learning exercises you did in class that worked well, along with determining which tools need improvement, will help you prepare and gauge your readiness to implement a new design.

If you hope to implement online videos of lectures and have found appropriate matches or made some of your own, test a few. Spread these trials out and carefully consider the material you choose to maximize the likelihood of a successful experience for you and your class. You can gradually infuse more of this approach over several semesters.

Assess what areas need the most work before full implementation. Do you have sufficient activities to do each class meeting for an entire semester? Are activities you may have found online or borrowed from colleagues working well? Perhaps they spark ideas for activities that would better fit your students and your course. The American Mathematical Society (AMS) Blogs website (https://blogs.ams.org/matheducation/) is a great place to search for additional answers as it provides a variety of articles offering insight into how others have solved problems in their classrooms. Most importantly, these are written by other mathematicians! The Mathematical Association of America (MAA) also offers some articles on teaching at their blog site, https://www.maa.org/community/maa-blogs.

Consider attending talks on the style of active learning in which you are interested and considering implementing. Being fully aware of the challenges and learning how others have overcome them will be an essential component of a successful and enjoyable experience. The Joint Meetings of the AMS and the MAA, as well as sectional meetings, include talks on educational topics. Attending events like these may also help you forge connections within the active learning community that may become excellent resources.

Finally, try to find a mentor or partner. Having a colleague with whom you can confer is invaluable any time you step outside your comfort zone. If you have the opportunity to work with a department member who has used the strategy, is currently exploring it, is interested in trying to do so in the future, or is simply supportive of your endeavor, you will have a significant asset as you undertake this initiative. If your campus has a teaching resource center, it should be able to offer support as you seek to educate yourself on new strategies, possibly connecting you with other faculty with similar interests. You may also consider reaching out to the author of an article or a speaker at an event who particularly inspired you or whose work you find intriguing.

Active Learning

I have discussed active learning quite broadly thus far, as arising from any activities or practices which have students thinking about, engaging in, and discovering material in class. The opportunities discussed to this point generally involved students predicting, attempting, practicing, or recalling content. As we move away from pure lecture and towards an enlightened guidance, there is a broad spectrum of levels of engagement that various approaches provide.

It has been found that benefits can be gained through a variety of degrees of these activities. The "Freeman Report" is a meta-analysis of 225 studies comparing the effects of traditional lecture versus active learning in undergraduate STEM courses. Though they allowed a spectrum of active learning tools and levels of intensity, they found improved test scores and decreased failure rates in the active learning classrooms (Freeman et al., 2014). The study included courses with only occasional activities such as group problem solving or in-class worksheets, supporting the notion that even small steps in this direction may have an impact on your students.

Benjamin Braun, Priscilla Bremser, Art Duval, Elise Lockwood, and Diana White wrote an informative series on active learning in the mathematics classroom as part of the 2015 AMS Blog, in which they suggest necessary considerations prior to attempting active learning, discuss some of the common challenges, and outline options for pursuing such exercises. I will explore some of their notions in this section.

Things to Consider Before Designing Active Learning Activities

Braun et al. note that the classroom and teaching environment, as well as the student and course goals, affect the implementation of active learning strategies (2015, September).

Class Size

Your class size affects many decisions throughout the semester. In Chapter 3, I mentioned that the method by which you get to know your students on the first day of class may be impacted by how many of them there are. If something as benign as taking attendance factors in the number of students present, it should be no surprise that the format of daily activities must consider and address this as well. Weigh your ability to field questions from students as they work independently or in groups during class. While undoubtedly some questions that arise will be appropriate for class discussion, there will also be individual mundane calculation errors to field. Will you be able to circulate through the class adequately to properly sort through and address questions? Your classroom may also have limitations. Be sure to inspect the room once it has been assigned to verify that it will support your course design. Request a room change if necessary and make the necessary modifications to your plans if your request cannot be accommodated.

Classroom Environment

If you are at a larger institution, you may not only face larger class sizes of hundreds of students but less overall class time. If your institution utilizes hybrid courses in which the course is split between online and in-person instruction, this will naturally affect how you will choose to spend the time in class and the active learning strategies you are able to employ (Braun et al., 2015, September). In this setting, you likely have little control over the institution's online content but may be permitted to supplement. The decreased in-person contact time and the style and content of the online portion will have implications for the activities you may do in class.

Teaching Environment

The teaching environment varies greatly among institutions. The time you have available for designing and preparing your courses is of utmost concern. If you have a heavy teaching load, high expectations regarding research or significant committee work, you may struggle to find the necessary time to devote to the proper planning and problem solving that designing and maintaining your course will take. Do you have support available to you at your institution, such as a mentor or pedagogical training? How dependent on performance evaluations is your position or standing at your institution? (Braun et al., 2015, September)

If you are a graduate teaching assistant or a new professor, you may be torn between engaging in intriguing innovation and establishing good references and student evaluations. This concern is fair and one reason I encourage you to take time to educate yourself and gain experience in the classroom, building up your classroom skills and repertoire of activities. The ability to create meaningful activities requires expertise (Braun et al., 2015, October 20).

Rushing into a new design may mean you lack the ability to offer your students the engaging tasks necessary to make the experience successful.

Course Role and Audience

The design of a course is impacted not only by the course content it is expected to deliver but also the role the course plays (general education, required or elective course in-major, etc.) and the students it typically attracts. The learning objectives you assign to your course will be affected by these facets and your students' goals and expectations will vary with courses as well (Braun et al., 2015, September). Consider whether your intended design will achieve results that align with the goals of your students and your institution.

Student Resistance

A final point to consider is that the more active your class design becomes, the more you may face resistance from students. In "student-centered" learning, there is a shift in responsibility; students take a more active role in class and might be given influence on course parameters. The professor aims to foster and guide students as they pursue the learning objectives for the course, rather than to simply state all the content directly. It is not uncommon for both the professor and the students in the class to have reservations or feel uncomfortable the first time they participate in a student-centered course (Felder & Brent, 1996). Students are often less than enthusiastic about this shift in pedagogy if they have not experienced it before. Just as it may be challenging for you to draw suggestions and ideas out of your students, rather than simply telling them the answer, your students may resent a shift away from being directly told everything that they need to know. Having an unreceptive or downright hostile audience is hardly an ideal setting when trying something new. You will likely be less confident than usual and your excitement at the prospect of a novel approach may wane pretty quickly. Take comfort in the reality that an initial discomfort on both sides is to be expected; but if done well, this process can lead to a deeper level of learning (Felder & Brent, 1996).

The Art of Telling

With varying degrees, active learning techniques scale back how much you tell your students directly. I like explaining mathematics; that is why I pursued this profession. My enthusiasm for a topic can make it challenging for me to hold back blatant exposition, but doing so gives students an important opportunity for a deeper level of learning. At some points, it is necessary and appropriate to tell students information directly. The aim is to avoid exposition when there is an opportunity for students to engage with the material (Braun et al., 2015, October 20).

The goal in active learning is to encourage students to discover solutions based on their own investigation, discussion, and thought process. The instructor remains effective by selecting engaging tasks, anticipating students' reasoning and strategies, fostering, and directing discussion without acting as the sole authority, and engaging in judicious telling (Smith, III, 1996: 397). Telling can be used to note contradictions in students' ideas, provoke alternate approaches to concepts, and foster students' explanations (Lobato, Clarke & Ellis, 2005: 108). Teachers can use telling to encourage students to make sense of a concept and elicit their mathematical ideas (Lobato, Clarke & Ellis, 2005: 110–11).

Telling is useful both in the setup and conclusion of an exercise in discovery. The most basic example of necessary telling is to define mathematical terminology. There is no concept for students to discover or with which they might engage; they simply need to hear or read a definition so that they may have a basis for a problem put before them. It is potentially unrealistic and counterproductive to ask students to develop new formulas, predict conventions, or discover concepts without a structured guidance in place. Additionally, it is important following an extended period of student engagement to provide a conclusion, clarifying, and confirming the discoveries that have been made in the session (Braun et al., 2015, October 20). Students may miss important aspects which they are better able to learn and signify from exposition after they have gone through the discovery period (Schwartz & Bransford, 1998). Wrap-up sessions can still encourage student involvement but any questions posed then should not be left open-ended indefinitely.

Careful observation and assessment of students' progress is necessary to determine what additional telling might be prudent. During a class discussion, you may need to provide clarification or further exposition following an exercise that involved group work or mobile/clicker responses (discussed further in the next section). This presents the material at a time when students are primed to receive it, due to the engagement in which they have just participated (Braun et al., 2015, October 20). This form of telling might be avoidable with further questioning and discussions between students, but it is more effective than the telling which occurs in a traditional lecture. This may be a useful strategy if your department has a set schedule for your course, as it allows you to include engaging opportunities for students and remain on-track.

Additional Methods to Achieve Active Learning

In *A Typical Class*, I discussed a number of active learning activities. The opening and closing activities ask students to recall and reflect on material which they have learned or make predictions about questions they do not yet have the ability to answer. Having students work problems independently, in pairs or small groups, engages students in the material as it unfolds, rather than after a lengthy lecture, and provides an opportunity for collaboration. The interactive in-class handouts and review sheets in Chapter 5 foster retrieval, engagement, and reflection. This section offers additional activities for active learning. For more options, consider the ample list provided by Sebastian M. Marotta and Jace Hargis (2011) in *Low-Threshold Active Teaching Methods for Mathematical Instruction*, as well as Jennifer L. Faust and Donald R. Paulson's (1998) *Active Learning in the College Classroom*.

Think-Pair-Share

A relatively simple avenue to active learning is the "think-pair-share" technique. In this activity, you ask students to work on a problem for a few minutes on their own, and then have them compare their answers with those sitting around them or with a designated partner. Students subsequently share their answers with the class as a whole or the nearby groups in class. This method allows students the opportunity to think about the material in class and explain their thinking to one another and gives instructors immediate insight into the students' level of comprehension (Braun et al., 2015, October 1).

Clickers or Mobile Polling

Clickers are student response systems with which students answer multiple-choice questions posed to the class. Mobile phones and laptops can also be used in this capacity by texting answers or participating in online polling. In the latter, the instructor receives data as to how many students responded with each answer. Not only are your students engaged in thinking about the material but you also receive precise real-time feedback as to the class' comprehension and the specific misconceptions that are occurring.

Clickers or mobile polling can be used as part of the "share" in the "think-pair-share" (Braun et al., 2015, October 1) or in a modified "think-share" scenario when pairing presents a challenge due to time constraints or an easily distracted group of students. Using devices in the latter case, which is absent of peer interaction, may diminish the level of engagement and increased learning. Former MAA president, David M. Bressoud, notes in his piece on the use of clickers, *"Should Students Be Allowed to Vote?"* that the true impact arises not from the devices themselves, but their ability to provide a structured setting for student interaction and to hold groups accountable (Bressoud, 2009).

Polling is an excellent way to place natural breaks in your lesson and ensure your students are thinking about the concepts rather than just jotting down notes. Bressoud claims that "[t]he problem is not that students don't want to think about the mathematics. It is that they do not know how. Clicker questions help structure these reflective moments …" (2009). Students become engaged with the material in a natural and productive way as you are teaching it. Having students process information in class and provide you with the necessary feedback on whether they are ready to advance or need additional discussion is clearly useful.

Writing good multiple-choice questions can be challenging, but fortunately there are readily available resources for using polling devices in the classroom. There are entire texts written for their use and free resources available online as well. Carroll College provides materials which were developed through two projects funded by the National Science Foundation (NSF), "MathQUEST: Math Questions to Engage Students" and "MathVote: Teaching Mathematics with Classroom Voting." They have questions available for a variety of topics on their site (http://mathquest.carroll.edu/) and a link to a list of additional resources including reading on the use of clickers and links to additional problem sets. Cornell also has a host of resources, including problem sets and editable questions in LaTeX, available at http://www.math.cornell.edu/~GoodQuestions/materials.html.

If your institution does not own clickers and you prefer not to have students use phones or laptops, you can use colored pieces of paper or large lettered notecards to conduct voting in the classroom. This avoids delays that technical problems can cause and the time-consuming process of getting started with a classroom response system. This is also a great way to test the waters of polling prior to incorporating it as a regular component of a course. One drawback to a low-tech approach is that you are not provided the instant summary data which a computerized system can offer.

Peer Review

In peer review, students critique and correct a partner's work. In the context of a computational exercise, students can not only verify their partner's accuracy but also whether ample supporting work was provided and whether it was easy to follow. For a more theoretical question or a proof, partners can check for valid arguments and proper form, conciseness, and level of completion. A challenge to using peer reviews is that students may attempt to give as positive a report as possible to their partners. While a constructive tone is important, ignoring incomplete, inaccurate, or weak answers defeats the purpose.

Maximize the effectiveness of peer reviews by purposefully demonstrating useful feedback. Have students reflect on the uselessness of receiving minimal criticism to help students see their critiques as constructive and designed to assist their partners and model the quality of feedback you would like to see in peer reviews in your own grading of students' work (Moore, 2016). Consider drafting a few sample problems that you can review together as a class. Ask students to take a moment to write up the errors or weaknesses of the proposed answer, and then have the class formulate a review of the work. Giving students a few examples may help them approach their peers' work more analytically.

An alternative to pairs is to have a group of students review an unknown student's work. For privacy purposes, you could select work from another term and present it in your own handwriting. Knowing that the author is not present will allow students to freely offer critique.

Group Evaluations

As with the peer reviews, students may evaluate and provide feedback for the work done by a group of students. If you have activities for which you will have students break into groups, you might consider student presentations of the results which are evaluated by the rest of the class. This can serve to provide necessary corrections and clarifications and to keep groups on-track if they are working on these problems during class.

Inquiry-Based Learning (IBL)

In inquiry-based learning, the instructor introduces a concept or skill, then poses a question that requires students to speculate, make discoveries, and reflect. This technique might be applied for select topics within your course, or you could consider an overall inquiry-based course design, discussed in the next section. As mentioned there, there are publicly available question sets for math courses at *The Journal of Inquiry-Based Learning in Mathematics* (http://jiblm.org), which you may find helpful.

In her article, *Turning Routine Exercises Into Activities that Teach Inquiry: A Practical Guide*, Suzanne Ingrid Doreé (2016) showcases opportunities to bring inquiry into your classroom. Small steps, such as rephrasing routine exercises as questions and asking students to make conjectures and find counterexamples, can be done long before you are ready to take on a full IBL course design. Using these activities should enhance learning and may entice you to pursue a more in-depth inquiry approach.

Just-in-Time Teaching (JiTT)

Developed by Gregor Novak, Andrew Gavrin, and Evelyn Patterson, this approach involves persistent formative assessments. Students are asked to read material and complete online assignments prior to class. These assignments serve to not only assess students' prior knowledge, but to engage students in thinking about the concepts to be discussed in class, better preparing them for deeper discussion. Because the professor receives the responses before class (i.e. "just in time"), the lesson can be modified accordingly to address deficiencies and/or utilize foundational information which is deemed to be in place. Visit the JiTT website https://jittdl.physics.iupui.edu/jitt/ for more information and sample problems. The Good Questions Project has adapted this method for calculus courses and provides problems for the "pre-class warmup" as well as in-class problems on the website http://pi.math.cornell.edu/~GoodQuestions/.

Course Designs for Active Learning

As I mentioned in the introduction to this book, the course designs described in this chapter are presented here purely for the reader's exposure. The descriptions provided should not be taken to be complete analyses of these designs, nor should you feel pressure to pursue them at this time. Each requires extensive preparation and would ideally be done initially with the guidance of a mentor. If you are intrigued by a course design described here, consider further study, and test the waters in small ways. Once you have a healthy collection of active learning techniques, you may decide you would like a more pervasive change in your overall course design. Here are some common designs which seek to enable a maximum level of student engagement and activity in the course.

Community Service Learning (CSL)

When students feel a task in your course is important, they may feel increased motivation (Pintrich, 2003: 674–5), so finding impactful projects in your immediate community may improve the enthusiasm with which your students approach the material. In *Using Sustainability to Incorporate Service-Learning Into a Mathematics Course: A Case Study*, Victor J. Donnay describes how students performed service to their own community through cost-benefit analyses of a variety of energy-saving initiatives on campus and in their township (2013). On-campus studies included the financial impact of eliminating trays in the dining halls, using conservation modes for heating and cooling buildings at night and using LED lightbulbs. An off-campus project involved working with the architectural team designing a local recreation center to estimate the impact of installing a geothermal energy system instead of a traditional heating system. This work was extended in a summer research project assisting town officials in the submission of a grant proposal for funding towards the geothermal system. Courses which demonstrate and utilize the skills of a course in this way are examples of community service learning.

The ability to provide a service to the community while engaging students by actively demonstrating the usefulness of the material they are learning is an appealing and refreshing twist on a typical course. Such rewards do not come for free, as these projects require extensive work to launch and maintain. In her article, *What Makes Service-Learning Unique: Reflection and Reciprocity*, Barbara Jacoby describes how the learning acquired through a service-learning course goes beyond the experience, with reflection that occurs throughout: "In service-learning, opportunities for learning aren't incidental to the course — instead, they are integrated into the course or program structure, instead of being added on at the end. Reflection must be designed by intention to facilitate the desired learning outcomes" (Jacoby, 2013). This suggests that you should not consider CSL as something you might just "add" to a course; rather, you will need to think of how to *redesign* your course. Jacoby outlined the principles for designing such a course in an online seminar, *Service-Learning Course Design: What Faculty Need to Know*. An abbreviated outline of her suggestions follows (Bart, 2010):

- Determine which learning objectives can be achieved through service learning and how.
- Determine how academic content will complement the service component to your course to ensure all learning outcomes are achieved. Determine the activities through which reflection will be achieved.
- Establish a community partnership and discuss the organization's needs.
- Determine your assessment methods and schedule. Determine any role the community organization may have in evaluating students.

- Create a syllabus which explains the rationale for utilizing service learning in your course, describes the expectations of students and their responsibilities, and describes the course activities and assessments. Review this syllabus with your community partner and revise as needed.
- Address any logistical or safety issues that may exist.

Even this skeletal list more than hints at the extensive thought and preparation one must complete. Clearly, such an undertaking should be done only when appropriate for a course's content and learning objectives. There are a variety of interesting opportunities as evidenced above.

Another option for CSL in mathematics is having students tutoring or engaging in educational exercises in primary or secondary schools. Ekaterina Yurasovskaya of Seattle University has used service learning in a precalculus course, having students spend two to three hours a week tutoring mathematics in the community, at mostly middle-school or higher levels. The motivation was to have students improve their prerequisite algebra skills by teaching them to younger students. In a November 2017 AMS blog, Yurasovskaya observed improvement in her CSL sections and noted a decrease in fundamental errors. She also found a qualitative difference in the atmosphere of these classes, noting that students seemed more focused and dedicated to their work. She does offer a buyer-beware warning that CSL demands significant time, may lead to lower course evaluations – even when students enjoy the community experience – and an instructor may not be able to control the experience (Yurasovskaya, 2017). This sentiment echoes Donnay's observations that students both appreciated the community service learning experience and felt uneasy with the unpredictability of the direction of their work (2013: 532). He offers the caution that CSL faculty need to come to terms with the discomfort they may feel due to the decreased control over the trajectory of the course.

This suggests that a professor taking an interest in a CSL course should choose a course that not only lends itself well to a community project but also is a course with which the professor has a high degree of confidence. Having comfort with the flow and structure of the course will be of assistance when unexpected events require a change. It may serve instructors well to consider selecting courses in which they have demonstrated a history of solid evaluations so that a semester or two of lower evaluations will not be of concern as the experiment unfolds and the project design matures. Additionally, selecting a semester with a lighter teaching load or less frequent committee meetings would be advantageous.

Inquiry-Based Learning (IBL)

IBL may be the most well-known example of active learning in mathematics (Braun et al., 2015, October 1). The Academy of Inquiry Based Learning defines IBL as "a broad range of empirically validated teaching methods which emphasize (a) deeply engaging students and (b) providing students with opportunities to authentically learn by collaborating with their peers" (2019). In the mathematical setting, IBL has been identified to have the following characteristics (Hotchkiss & Fleron, 2014):

- Coursework is primarily problem solving.
- Class time is comprised of student-centered activities, such as group work and presenting and critiquing problems at the board.
- The exposition of material is achieved mainly through student investigation.
- The professor's role is decentralized. The instructor seeks to guide and support rather than directly state content.
- Students actively navigate how class is spent and take increased responsibility for communication and learning.
- Students utilize reflection and oral and written communication to gather and express new ideas, learning strategies, and mathematical ideas.

A sample class might involve a short mini-lecture or introduction to the day's topic of exploration, followed by students working together to investigate a problem then presenting and discussing problems. Alternatively, it might start with students presenting problems previously completed, followed by the class reviewing and critiquing the work presented.

An IBL course strategy will involve significant start-up work. Consider reading through course materials that have already been prepared, such as those provided by *The Journal of Inquiry-Based Learning in Mathematics* (http://jiblm.org). Looking at some sample lessons will provide insight as to how the overall IBL process works, in addition to offering exercises you may be able to use. If you teach a course which you would like to transition to IBL in the future, consider trying out a few of the problems or activities to get a sense of how students respond to them. You may discover you need to tweak these to better fit your audience or offer additional setup of questions. Reflect on which aspects of the session worked well. Continue to experiment in small ways until you find your footing and pursue additional education and training in IBL to address concerns.

Flipped Classrooms

In a flipped (or inverted) classroom design, the content material is presented outside of class, typically in the form of assigned reading or videos of lecture. The time in the classroom is spent with active learning exercises, discussion of topics, and student questions. In a mathematics class, videos are frequently implemented, as assigned reading in a standard text may not be sufficient.

Michelle Estes, Rich Ingram, and Juhong Christie Liu propose a three-stage design for a flipped classroom, based on their extensive review of the research literature related to flipped learning (2014). A pre-class stage allows for pre-assessment and initial instruction, and the in-class stage clarifies concepts through discussion, instructor questions, and possibly peer instruction or review. The post-class stage can be used for assessment of learning and to apply or transfer the knowledge attained. The authors acknowledged challenges to the flipped classroom, including the need for students to be capable of identifying and demonstrating self-directed learning skills but concluded that there can be several significant benefits to this design. They cite potential benefits of optimizing class time, developing thinking skills and enhancing the interactions in the classroom.

One potential challenge noted by Estes's team was the creation of media for the pre-class instruction. An ample supply of online videos are available for your use. Khan Academy and MIT Open Courseware are two examples of free online videos of mathematics lectures available. Smaller collections can be found on the websites of major institutions who have implemented flipped courses. University of Hartford has posted the materials developed to flip the Calculus I course offered there. The site (http://math.hartford.edu/flipping/index.html) provides videos, in-class activities, and quizzes. It also contains videos for a Calculus II course. Ohio State University has posted online lectures for a handful of courses, including two semesters of calculus at https://mslc.osu.edu/online/lessons. Additionally, if you are using an online text and homework program, there may be videos available within the online access. While using readily available lectures can save you time and offer your students variety in the instruction they receive, be sure to watch these through to note the coverage and notation that is used. The in-class discussion should be consistent, or at least highly compatible, with the material students have been asked to watch or read outside of class.

Using your own videos maintains a seamless transition between what is presented online and how you handle the material in class, but may seem too time consuming to set up, especially if you just want to test the waters of this approach. Your institution may offer a service of video recording lectures as a tool for you to use in assessing the effectiveness of your teaching. You could get double duty out of such a recording, if permissible by your institution, but you will need to be mindful of protecting the privacy of all students in the class recorded if you are posting it online. Considerations for crafting your own videos is discussed later in this chapter.

Balance your reliance on available materials with your own input. It makes sense to utilize resources that have been tested in classrooms and tweaked for improvement, but consider your students' perspective if you run an entire course using materials from another institution. Might they question your influence on their learning? Take time to tailor the resources to fit your course as you would like it. Adjust the content as needed, inject alternative videos when the ones posted are not in line with how you feel the material is best presented, and be an energizing presence in the classroom.

I mentioned earlier that students may be resistant to a course which is highly student-centered. One student described dissatisfaction with a flipped classroom experience in an article in *The Mercury News*, stating a preference for direct exposition to ensure student comprehension (The Life in Perspective Board, 2014). In the article, the student describes a teacher experimenting with a flipped classroom for a multivariable calculus course in a college-prep school, using MIT Open Courseware videos to deliver the course content prior to class time. The experiment reportedly went well the first week but students began to chat during the second week instead of completing the in-class assignments. The student complained that the videos required significant time to watch, especially when encountering a confusing point which required rewinding and re-watching. The teacher abandoned the experiment, clearly leaving the student, who advised teachers against flipping a classroom, with the impression that this approach could not work, saying "[y]our students will have difficulty with the concepts, and there won't be enough of you to go around and answer all of their questions. After all, the 'normal' classroom has worked since the beginning of standardized education" (The Life in Perspective Board, 2014).

I do not mention the potential wariness of students to dampen any enthusiasm you may be feeling towards trying this approach, but rather to encourage you to fully prepare for such an endeavor. In the above example, promoting more on-task group discussion and alternating moments of discussion as an entire class may have helped with addressing more student questions. Using shorter video segments could have aided the students' ability to digest the information.

Estes, Ingram, and Liu specifically note that success in the flipped classroom depends upon positive interactions between the instructor and students and continued motivation and contribution throughout the process (2014). The more classroom experience you have, the better you may be able to navigate challenges. Consider waiting to try a design such as flipping a classroom until you have successfully implemented a number of active learning exercises in your classes for several semesters and learned how to address student resistance (Felder & Brent, 2016: 144).

At least one study found no significant difference between the flipped classroom design and that of an active learning classroom which required similar learning processes (Jensen, Kummer & Godoy, 2015). Students participated in engagement and exploration, and heard explanations from the professor before class in the flipped setting versus during class in the non-flipped sections. The flipped sections then completed elaboration (extending content to novel settings) and evaluation (assessments) during class, while the non-flipped section did so after class. Both course designs used technology for the portion completed outside of class. The researchers cited the essential component of both approaches as active student engagement and constructive approach to discovery of content.

Flipped Flipped Classrooms

In a flipped flipped classroom, material is still presented outside of class and class time is used for exploration but the order is reversed. Students do exploratory work in class and then follow the class discussion with online presentations or reading. The challenges with student engagement and acceptance of the style will persist in this format but this design may feel less stressful to students since the initial exposure to material happens in your presence and with their classmates. The knowledge that they will have an opportunity to seek out more answers after class through the readings or video lectures may decrease anxiety they might feel with confusion about a concept in class. The theory is that they have been better primed for the clarification on points of confusion and they will actively seek answers to questions that came up in class.

Online Courses

As with the course designs discussed in the previous section, a great deal of work goes into creating an effective online course. There is more to be done than simply typing up lectures used in a traditional class and posting them online. Both the initial setup of the course and the online interaction are time-consuming. Student engagement is necessary, as is the ability of the instructor to gauge the students' level of comprehension.

Online courses should involve interaction with the professor and students should have opportunities to interact with one another (Englebrecht & Harding, 2005: 257–8). The necessary communication for these conversations brings its own challenges. If communication is done in real time, such as via an online chat session, then some of the liberating advantages of the online experience are lost. Students and instructor must all be available at the same time and the pace of the discussion may be too fast or too slow for some learners. If the communication is done through posts which can be responded to later, the feedback is delayed and student engagement may be decreased (Englebrecht & Harding, 2005: 267–8). In comparison to a face-to-face class, students in either real-time or delayed discussions may produce longer and more thoughtful comments and a greater percentage of the class may get involved (Brewer & Brewer, 2015: 35). The course design should implement a blend of styles of communication which maximizes the collaboration of those involved.

One challenge with online courses is assessment. Security is a concern if assessment is done entirely online, though there are options available to monitor students remotely. A hybrid approach, in which students take exams in person, provides security, but potentially poses geographic restraints on those enrolled, may hinder those with mobility issues, or may carry additional costs if testing centers are used.

Creating Videos for Online Courses

Videos can provide a significantly more engaging environment than typed-up lectures, though they likely require more time to watch than reading a lesson. Before attempting to create videos, consider watching a variety of those available online, such as those mentioned in the section *Flipped Classrooms*. This will expose you to different approaches to presenting the material and help you gain a sense of how you might like to go about your own exposition. It is not recommended to simply record your usual 50-minute lecture for a topic and post it online. Instead, consider the following recommendations.

Philip J. Guo, Juho Kim, and Rob Rubin performed a large study on how the designs of videos used in xMOOCs (Massive Open Online Courses, with a central professor) related to student engagement. While the overall course design and structure of a MOOC, which generally has little to no in-person contact with a professor, may differ greatly from that of a traditional lecture course or a flipped classroom, this study is of interest since its focus was engagement in online videos. Of additional importance is that the study involved four *math and science-related courses* (Introduction to Computer Science and Programming, Statistics for Public Health, Artificial Intelligence, and Solid State Chemistry) and examined data from *6.9 million* video watching sessions. The researchers measured engagement in two ways, engagement time (the length of time that a student spent watching a video) and problem attempt (whether a student attempted the follow-up problems, available after 32% of videos, within 30 minutes after watching the video). Due to the scale of the study, Guo and his colleagues acknowledge that they were unable to measure "true engagement," the degree to which a student was paying attention to the video, since this requires direct observation and questioning (Guo, Kim & Rubin, 2014). The quantitative data collected was supplemented with qualitative information from six staffers involved in producing the courses.

Guo, Kim, and Rubin found that the following styles of videos are more engaging:

- shorter videos
- talking-head videos (favored over those with just slides)
- videos with a personal feel
- videos utilizing Khan-style tablet drawings (favored over slides)
- videos with fast-speaking professors, with high enthusiasm

Guo and his colleagues note that the speed of a professor's speech was often tied to the enthusiasm demonstrated. Thus, this finding does not suggest that one should strive to speak quickly in a lecture, but rather that one should not arbitrarily slow speech. To achieve optimal engagement, their recommendations include:

- significant pre-production lesson planning to craft 6-minute video sessions
- post-production editing to insert video of the instructor's head at appropriate times
- using an informal setting
- utilizing motion and continuous visual flow, along with dialog
- if recording a traditional lecture, planning it as a series of short, discrete mini-lectures that can be edited later for online posting
- being enthusiastic in video and not slowing your speech unnaturally

If you want to use your own lectures but have limited preparation time, consider the suggestion of breaking a typical lecture into subtopics so that you can quickly edit the video into separate components. The inability for students to clarify points in your lecture by asking questions means that a longer online lecture may be less effective. Additionally, when a recorded lecture is on a single subtopic, it is easier for students to replay videos on specific points of concern.

A Final Note on Innovation

It can be tempting both to do what feels natural and to explore something new and exciting. You may feel pressure to further your career by securing solid evaluations and by being innovative. These achievements may not necessarily go hand-in-hand when you are in the early stages of a new methodology. It is okay to rely on what you feel you already do well and also strive for what you feel you *might* be able to do well.

You do not have to fully embrace or employ an approach to utilize some of its strategies. Braun and his colleagues note that it is important to acknowledge a range of active learning techniques are effective and teachers need not transition to all-student-discovery to pursue their desire to improve learning and engagement (Braun et al., 2015, October 1). In his book, *How the Brain Learns*, David Sousa takes the above idea farther, stating "... neither lecture nor any other method, for that matter, should be used almost all the time. Successful teachers use a variety of methods, keeping in mind that students are more likely to retain and achieve whenever they are actively engaged in the learning" (2011: 102).

One reason not to force a methodology simply because "research shows" it is the best way for students to learn is that the data cannot tell the full story of the necessary components for success. The authors of the Freeman Report note that the instructors in the studies analyzed voluntarily implemented active learning techniques and "[i]t is an open question whether student performance would increase as much if all faculty were required to implement active learning approaches" (Freeman et al., 2014). The professors using newer active learning techniques and publishing their results and analyses are those who are keenly interested in experimenting with, and pushing the boundaries of, their teaching. This is a specific sub-population of mathematics faculty as a whole. These faculty members may be at more supportive institutions, have had more access to training and feedback than average, have mentors guiding and directing their efforts, and/or may have security in their positions (i.e. tenure). Their own original pedagogical backgrounds and methodologies may have been more adaptable to the new strategies than the typical professor and they were likely to be highly motivated to address failures and adapt their strategies. These faculty members may be especially talented at reading their students, motivating discussion, or exciting the class. My point is not to suggest only a rare few are capable of success, but to alert you to many factors that can influence success. Pursue ideas that excite you with thorough preparation and elicit guidance from those who have attempted similar strategies.

Conclusion

Mathematics and its instruction share many parallels. Each may bring great highs and lows. From time to time, you will likely fail to achieve the learning experience you most wanted to create. A carefully crafted exercise may not illuminate, entice, or engage. You may have beautiful moments when everything goes perfectly or when you witness a student's epiphany. A single student's hard-earned success may bring you surprising restoration and rejuvenation.

Like the subject itself, teaching mathematics requires intuition and close attention to detail and necessitates a great deal of reflection, practice, and resilience. Just as you may have labored to master a particular skill or concept in mathematics, you may find that it takes some time to get your proper footing in the classroom. Math is not always easy, but that is when it is interesting. Without struggle, there is no thrill or glory in finding a solution. The same is true in teaching.

There is hardly ever just one right way to solve a problem in math or in the classroom. That is part of the beauty of the field and the profession. Whether you are embarking on a lifelong career or just visiting for a bit, I hope you will enjoy the challenges and creative solutions ahead.

References

The Academy of Inquiry Based Learning [online]. (2019). Available at: http://www.inquirybasedlearning.org/. Accessed October 20, 2019.

Ambrose, S. A., Bridges, M. W., DiPietro, M., Lovett, M. C. & Norman, M. K. (2010). *How learning works: Seven research-based principles for smart teaching*. San Francisco: Jossey-Bass.

Badger, M. S., Sanguin, C. J., Hawkes, T. O., Burn, R. P., Mason, J. & Pope S. (2012). *Teaching problem-solving in undergraduate mathematics*. [pdf] Coventry, England: Coventry University. Available at: http://www.mathcentre.ac.uk/resources/uploaded/guide.pdf. Accessed November 24, 2019.

Bain, K. (2004). *What the best college teachers do*. Cambridge: Harvard University Press.

Bart, M. (2010). Six steps to designing effective service-learning courses. *Faculty Focus*, [online] April 21. Available at: https://www.facultyfocus.com/articles/curriculum-development/six-steps-to-designing-effective-service-learning-courses/. Accessed October 20, 2019.

Boaler, J. (2016). *Mathematical mindsets: Unleashing students' potential through creative math, inspiring messages and innovative teaching*. San Francisco: Jossey-Bass.

Boice, R. (2000). *Advice for new faculty members*. Needham Heights: Allyn Bacon.

Braun, B., Bremser, P., Duval, A., Lockwood, E. & White, D. (2015). Active learning in mathematics, part I: The challenge of defining active learning. *AMS Blogs*, [online] September 10. Available at: https://blogs.ams.org/matheducation/2015/09/10/active-learning-in-mathematics-part-i-the-challenge-of-defining-active-learning/. Accessed July 4, 2018.

Braun, B., Bremser P., Duval, A., Lockwood, E. & White, D. (2015). Active learning in mathematics, part III: Teaching techniques and environments. *AMS Blogs*, [online] October 1. Available at: https://blogs.ams.org/matheducation/2015/10/01/active-learning-in-mathematics-part-iii-teaching-techniques-and-environments/. Accessed July 4, 2018.

Braun, B., Bremser P., Duval A., Lockwood, E. & White, D. (2015). Active learning in mathematics, part V: The role of "telling" in active learning. *AMS Blogs*, [online] October 20. Available at: https://blogs.ams.org/matheducation/2015/10/20/active-learning-in-mathematics-part-v-the-role-of-telling-in-active-learning/. Accessed July 4, 2018.

Bressoud, D. M. (2009). Should students be allowed to vote? *MAA Launchings*, [online] March. Available at: https://www.maa.org/external_archive/columns/launchings/launchings_03_09.html. Accessed July 4, 2018.

Brewer, P. E. & Brewer, E. C. (2015). Pedagogical perspectives for the online education skeptic. *Journal on Excellence in College Teaching*, 26(1), 29–52.

Brown, P. C., Roediger, H. L. & McDaniel, M. A. (2014). *Make it stick: The science of successful learning*. Cambridge: Harvard University Press.

Burnham, J. J., Hooper, L. M. & Wright, V. H. (2012). Top 10 strategies for preparing the annual tenure and promotion dossier. *Faculty Focus*, [online] April 25. Available at: https://www.facultyfocus.com/articles/faculty-evaluation/top-10-strategies-for-preparing-the-annual-tenure-and-promotion-dossier/. Accessed July 4, 2018.

Carter, S. P., Greenberg, K. & Walker, M. S. (2016). The impact of computer usage on academic performance: Evidence from a randomized trial at the United States Military Academy. *Economics of Education Review*, 56, 118–132.

Case, K., Bartsch, R., McEnery, L., Hall, S., Hermann, A. & Foster, D. (2008). Establishing a comfortable classroom from day one: Student perceptions of the reciprocal interview. *College Teaching*, 56(4), 210–214. doi: 10.3200/CTCH.56.4.210-214.

Cashin, W. E. (1995a). Answering and asking questions. *IDEA Paper No. 31*. Manhattan, KS: Center for Faculty Evaluation and Development, Kansas State University, January.

Cashin, W. E. (1995b). Student ratings of teaching: The research revisited. *IDEA Paper No. 32*. Manhattan, KS: Center for Faculty Evaluation and Development, Kansas State University, September.

Cavanagh, S. R. (2016). *The spark of learning: Energizing the college classroom with the science of emotion*. Morgantown: West Virginia University Press.

Chappell, K. K. (2006). Effects of concept-based instruction on calculus students' acquisition of conceptual understanding and procedural skill. In: F. Hitt, G. Harel, A. Selden, eds., *Research in collegiate mathematics education VI, part of the CBMS Issues in mathematical education series*, 13. Providence: American Mathematical Society, 27–60.

Chappell, K. K. & Killpatrick, K. (2003). Effects of concept-based instruction on students' conceptual understanding and procedural knowledge of calculus. *PRIMUS*, 13(1), 17–37. doi: 10.1080/10511970308984043.

Donnay, V. (2013). Using sustainability to incorporate service-learning into a mathematics course: A case study. *PRIMUS*, 23(6), 519–537. doi: 10.1080/10511970.2012.753649.

Doreé, S. I. (2016). Turning routine exercises into activities that teach inquiry: A practical guide. *PRIMUS*, 27(2), 179–188. doi: 10.1080/10511970.2016.1143900.

Elbow, P. (1986). *Embracing contraries: Explorations in learning and teaching*. New York: Oxford University Press.

Englebrecht, J. & Harding, A. (2005). Teaching undergraduate mathematics on the internet. *Educational Studies in Mathematics: An International Journal*, 58(2), 253–276. doi: 10.1007/s10649-005-6457-2.

Estes, M. D., Ingram, R. & Liu, J. C. (2014). A review of flipped classroom research, practice, and technologies. *International Higher Education Teaching & Learning Review*, [online] Volume 4, Article 7, July 29. Available at: https://www.hetl.org/a-review-of-flipped-classroom-research-practice-and-technologies/. Accessed October 20, 2019.

Faust, J. L. & Paulson, D. R. (1998). Active learning in the college classroom. *Journal on Excellence in College Teaching*, 9(2), 3–24.

Felder, R. M. & Brent, R. (1996). Navigating the bumpy road to student-centered instruction. *College Teaching*, 44(2), 43–47. doi: 10.1080/87567555.1996.9933425.

Felder, R. M. & Brent, R. (2008). Student ratings of teaching: Myths, facts, and good practices. *Chemical Engineering Education*, 42(1), 33–34.

Felder, R. M. & Brent, R. (2016). *Teaching and learning STEM: A practical guide*. San Francisco: Jossey-Bass.

Fink, L. D. (2003). *Creating significant learning experiences: An integrated approach to designing college courses*. San Francisco: Jossey-Bass.

Fink, L. D. (2010). A self-directed guide to designing courses for significant learning. [online]. Available at: http://www.designlearning.org/wp-content/uploads/2010/03/Self-Directed-Guide.2.pdf. Accessed November 24, 2019.

Freeman, S., Eddy, S. L., McDonough, M., Smith, M. K., Okoroafor, N., Jordt, H. & Wenderoth, M. P. (2014). Active learning increases student performance in science, engineering, and mathematics. *Proceedings of the National Academy of Sciences*, [online] June 10, 111(23), 8410–8415. doi: 10.1073/pnas.1319030111. Available at: https://www.pnas.org/content/111/23/8410. Accessed September 2, 2019.

Fukawa-Connelly, T., Weber, K. & Mejía-Ramos, J. P. (2017). Informal content and student note-taking in advanced mathematics classes. *Journal for Research in Mathematics Education*, 48(5), 567–579.

Guo, P. J., Kim, J. & Rubin, R. (2014). How video production affects student engagement: An empirical study of MOOC videos. *L@S '14 Proceedings of the first ACM Conference on Learning@Scale*, [online] 41–50. Available at: http://up.csail.mit.edu/other-pubs/las2014-pguo-engagement.pdf. Accessed July 4, 2018.

Harel, G. (2013). Intellectual need. In: K. R. Leatham, ed., *Vital directions for mathematics education research*, New York: Springer, 119–151. doi: 10.1007/978-1-4614-6977-3_6.

Hembrooke, H. & Gay, G. (2003). The laptop and the lecture: The effects of multitasking in learning environments. *Journal of Computing in Higher Education*, 15(1), 46–64.

Hotchkiss, P. & Fleron, J. (2014). What is inquiry-based learning? *Discovering the Art of Mathematics: Mathematical Inquiry in the Liberal Arts*, [online] July 2. Available at: https://www.artofmathematics.org/blogs/jfleron/what-is-inquiry-based-learning. Accessed July 4, 2018.

Jacoby, B. (2013). What makes service-learning unique: Reflection and reciprocity. *Faculty Focus*, [online] November 1. Available at: https://www.facultyfocus.com/articles/curriculum-development/what-makes-service-learning-unique-reflection-and-reciprocity/. Accessed July 11, 2018.

Jensen, J. L., Kummer, T. A. & Godoy, P. D. M. (2015). Improvements from a flipped classroom may simply be the fruits of active learning. *CBE-Life Sciences Education*, 14, 1–12.

Kraushaar, J. M. & Novak, D. C. (2010). Examining the affects of student multitasking with laptops during the lecture. *Journal of Information Systems Education*, 21(2), 241–251.

Lang, J. M. (2013). *Cheating lessons: Learning from academic dishonesty*. Cambridge: Harvard University Press.

Lang, J. M. (2016). *Small teaching: Everyday lessons from the science of learning*. Hoboken: John Wiley & Sons, Incorporated. Available from: ProQuest Ebook Central. Accessed November 11, 2019.

Lee, S., Kim, M. W., McDonough, I. M., Mendoza, J. S. & Kim, M. S. (2017). The effects of cell phone use and emotion-regulation style on college students' learning. *Applied Cognitive Psychology*, 31, 360–366.

Lew, K., Fukawa-Connelly, T. P., Mejía -Ramos, J. P. & Weber, K. (2016). Lectures in advanced mathematics: Why students might not understand what the mathematics professor is trying to convey. *Journal for Research in Mathematics Education*, 47(2), 162–198. doi: 10.5951/jresematheduc.47.2.0162.

The Life in Perspective Board. (2014). Teen life: Why experiment with 'flipped classroom' failed. The Mercury News, [online] February 6. Available at: http://www.mercurynews.com/2014/02/06/teen-life-why-experiment-with-flipped-classroom-failed/. Accessed July 4, 2018.

Lobato, J., Clarke, D. & Ellis, A. B. (2005). Initiating and eliciting in teaching: A reformulation of telling. *Journal for Research in Mathematics Education*, 36(2), 101–136.

Luo, L., Kiewra, K. A., Flanigan, A. E. & Peteranetz, M. S. (2018). Laptop versus longhand note taking: Effects on lecture notes and achievement. *Instructional Science*, 46, 947–971. https://doi.org/10.1007/s11251-018-9458-0.

Marotta, S. M. & Hargis, J. (2011). Low-threshold active teaching methods for mathematical instruction. *PRIMUS*, 21(4), 377–392. doi: 10.1080/10511971003754135.

Mathematical Association of America. (2018). *MAA Instructional Practices Guide*. [online]. Available at: https://www.maa.org/sites/default/files/InstructPracGuide_web.pdf. Assessed on July 21, 2019.

Moore, C. (2016). Frame your feedback: Making peer review work in class. *Faculty Focus*, [online] June 6. Available at: https://www.facultyfocus.com/articles/teaching-and-learning/frame-feedback-making-peer-review-work-class/. Accessed July 12, 2018.

Mueller, P. A. & Oppenheimer, D. M. (2014). The pen is mightier than the keyboard: Advantages of longhand over laptop note taking. *Psychological Science*, 25, 1159–1168. doi: 10.1177/0956797614524581.

Murawska, J. M. & Nabb, K. A. (2015). Corvettes, curve fitting, and calculus. *Mathematics Teacher*, 109(2), 128–135.

Nabb, K. & Murawska, J. (2019). Motivating calculus through a single question. *PRIMUS*, 29(10), 1140–1153. doi: 10.1080/10511970.2018.1490360.

National Research Council. (2003). *Evaluating and improving undergraduate teaching in science, technology, engineering, and mathematics*. Committee on Recognizing, Evaluating, Rewarding, and Developing Excellence in Teaching of Undergraduate Science, Mathematics, Engineering, and Technology, M. A. Fox & N. Hackerman, eds., Center for Education, Division of Behavioral and Social Sciences and Education. Washington, D.C.: The National Academies Press.

Pintrich, P. R. (2003). A motivational science perspective on the role of student motivation in learning and teaching contexts. *Journal of Educational Psychology*, 95(4), 667–686.

Pólya, G. (1965). *Mathematical discovery: On understanding, learning, and teaching problem solving, II*. New York: John Wiley & Sons, Inc.

Rasmussen, C., Marrongelle, K., Kwon, O. & Hodge, A. (2017). Four goals for instructors using inquiry-based learning. *Notices of the AMS*, 64(11), 1308–1311.

Rosen C. (2008). The myth of multitasking. *The New Atlantis*, [online] 20, Spring, 105–110. Available at: http://www.thenewatlantis.com/publications/the-myth-of-multitasking. Accessed July 4, 2018.

Sana, F., Weston, T. & Cepeda, N. J. (2013). Laptop multitasking hinders classroom learning for both users and nearby peers. *Computers & Education*, 62, 24–31.

Schwartz, D. L. & Bransford, J. D. (1998). A time for telling. *Cognition and Instruction*, 16(4), 475–522. Available at: http://www.public.asu.edu/~mtchi/papers/SchwartzBransford1998.pdf. Accessed September 8, 2019.

Smith, III J. P. (1996). Efficacy and teaching mathematics by telling: A challenge for reform. *Journal for Research in Mathematics Education*, 27(4), 387–402.

Sousa, D. A. (2011). *How the brain learns*. 4th ed. Thousand Oaks: Corwin.

Steele, D. F. & Arth, A. A. (1998). Lowering anxiety in the math curriculum. *Education Digest*, 63(7), 18–23.

Stearns Center for Teaching and Learning. (2017). Documenting your teaching [online]. Available at: https://stearnscenter.gmu.edu/knowledge-center/general-teaching-resources/documenting-your-teaching/. Accessed October 20, 2019.

Sun, K. L. (2018). The role of mathematics teaching in fostering student growth mindset. *Journal for Research in Mathematics Education*, 49(3), 330–355.

Weimer, M. (2011). Revisiting extra credit policies. *Faculty Focus*, [online] July 20. Available at: https://www.facultyfocus.com/articles/teaching-professor-blog/revisiting-extra-credit-policies/. Accessed July 12, 2018.

Weimer, M. (2011). So what did we learn about extra credit? *Faculty Focus*, [online] August 11. Available at: https://www.facultyfocus.com/articles/teaching-professor-blog/so-what-did-we-learn-about-extra-credit/. Accessed July 12, 2018.

Whitley, B. E., Jr. & Keith-Spiegel, P. (2002). *Academic dishonesty: An educator's guide*. Mahwah: Lawrence Erlbaum Associates, Inc., Publishers.

Wiggins, G. & McTighe, J. (2005). *Understanding by design, expanded*. 2nd ed. Alexandria: Association for Supervision and Curriculum Development. Available from: ProQuest Ebook Central. Accessed May 14, 2020.

Wood, E., Zivcakova, L., Gentile, P., Archer, K., De Pasquale, D. & Nosko, A. (2012). Examining the impact of off-task multitasking with technology on real-time classroom learning. *Computers & Education*, 58, 365–374.

Yeager, D. S. & Dweck, C. S. (2012). Mindsets that promote resilience: When students believe that personal characteristics can be developed. *Educational Psychologist*, 47(4), 302–314.

Yurasovskaya, E. (2017). Learning by teaching: Service-learning in a precalculus classroom. *AMS Blogs*, [online] November 27. Available at: https://blogs.ams.org/matheducation/2017/11/27/learning-by-teaching-service-learning-in-a-precalculus-classroom/. Accessed July 4, 2018.

Index

academic dishonesty 38–9, 82, 89; addressing 106–7
academic honesty: through assessment 88–9; through environment 82
active learning 67, 73, 181, 197–8; considerations before implementation 181–3; course designs 188–94; methods for 185–8; in regard to telling 183–4
algebra skills 113–15
approachability 78–9
arc length 145–9
assessment 13–16, **16**; affect on academic honesty 88–9; exams 97–105; formative 69, 87; forms of 19–23, **21–3**; homework used as 90; quizzes 90–2; summative 99; weighting forms of 23–6, **24–5**
assignments *see* homework
attendance 26–8, **28**

backward design 15–16, **16**

calculators 32
calculus: differential 126–8; integral 134–6; sophomore 144–52
cheating *see* academic dishonesty
class meeting 64–70
classroom response system *see* polling
classroom strategy 15
clickers *see* polling
closing exercise 68; examples of 121–2, 128, 135–6, 152, 156–8, 159
community service learning 188–90

complex variables 160
conduct 28, 35–6, 39, 45, 50–1
connections: use of 12, 52, 60, 66; *see also* parallels
Corvette problem 75
course goals 7, **16**
course policies 18; overview 36
CSL *see* community service learning
curve-sketching 127

derivation: opportunities for 127–8, 135, 145–9
desired learning outcomes *see* learning outcomes
discussion sections 111–12
divergent questions 57
documentation: pedagogical education 178; peer evaluations 172; self-evaluation 169–70; *see also* worksheets
documentation worksheets *see* worksheets

encouragement 76–8
enthusiasm 74–6
evaluation: peer 170–3; self- 163–70; student 173–7; *see also* worksheets
evaluation worksheets *see* worksheets
exams **21**; final 99; grading 102–3; post-mortems 103–5; preparing students for 92–6; too-long 100–2; writing 97–100; *see also* review guides
exit paper 68; *see also* minute paper

expert blind spot 54
extra credit 108–10

flipped classroom 192–4
flipped flipped classroom 194
formative assessment *see* assessment
Freeman Report 181, 197
Fundamental Theorem of Calculus 75, 135

grades: calculation 23–5; composition 18–23, 26–8; course 107–10; *see also* assessment
group evaluations 187

handouts: graphing techniques 128–34; integration strategy 136–50; linear algebra 153; proofs 156; sophomore calculus 152; trigonometry 122–5
homework 20–1, **22**; as assessment 90; as practice 71
honor violations *see* academic dishonesty

IBL *see* inquiry-based learning
inquiry-based learning 187, 191
interleaving 68
Intermediate Value Theorem 75
inverted classroom *see* flipped classroom

JiTT *see* just-in-time teaching
just-in-time teaching 188

late papers 30–2
L. Dee Fink's Taxonomy 10–13, **11–12**
learning objectives 7–13, **16**
learning outcomes 7–13, **16**, 54–5
lesson writing 56–62
limits 61, 75, 126, 160
linear algebra 152–6

make-ups 30–2
massive open online courses 195
Mean Value Theorem 75, 127, 149
mid-term evaluations 174–5
mindset: fixed 76; growth 50–1, 76; instruction 76
minute paper 65, 68–9, 89, 174
mistakes 82–4
mobile polling *see* polling
MOOCs *see* massive open online courses
motivation 40, 74–6, 82

open-ended questions 57
opening exercises 65; examples of 121–2, 128, 135–6, 152, 155–8, 160

pace 72–4
parallels: examples of 116–7, 119, 151–2, 155–6, 160; use of 12, 52, 60, 66
parametric equations 151
partial derivatives 151
participation 21, **23**, 26–8, **28**
peer review 186–7; *see also* evaluation
phones 32, 34–5, 65, 185–6
polar coordinates 151, 160
polling 65, 69, 185–6
polynomial division 116–7, 121–2
practice 20, 68–9, 91; importance of 70–2
precalculus 115–22
predictions **12**; 65–6
preparation time 62–3
prerequisite quiz 51–2
projects 19–20, **22**, 26; community service **11**, 188–90
proofs: courses 156–9; opportunities for 127–8, 135

quick glance overview: class 85; course policies 36; exam writing 99; first day 53; late policies 32; learning outcomes 13; make-up policies 32; office hours 85; post-class notes 165; post-course notes 167; preparation 63; scheduling 30; terminology 16; typical day 69–70
quizzes 19–20, **22**, 90–2; *see also* prerequisite quiz

rapport 79–81
rational expressions 116–19
reflection: post-class 164–5; post-course 166–9
reflection worksheet *see* worksheets
retrieval 34, 65, 68–9
review guides 92–3, 136–44
review sessions 94–6
Rolle's Theorem 10, 127

schedule 29–30
self-evaluation *see* evaluation
sequences 152, 160
series 152, 156, 160
student-centered instruction 183, 191, 193
student evaluations *see* evaluation
study guides *see* review guides

206 Index

surface area 145–6, 149
summative assessment *see* assessment
syllabus 17, 36–41; promising 41; quiz 41, 51; review 50–1; sample 41–6

telling 183–4
think-pair-share 185

vector space 155
video lectures: available 192; creating 185–7

wait time 74
worksheets: mid-term evaluation 175; peer review 172–3; post-class reflection 165; post-course reflection 168–9; student evaluation summary 177; *see also* handouts

xMOOCs *see* massive open online courses

Printed in the United States
By Bookmasters